SOIL and WATER RESOURCES:

Research Priorities for the Nation

These proceedings represent the best collective judgment on research priorities by a group of informed and dedicated people concerned with the future role of soil and water resources in meeting human needs

SOIL and WATER RESOURCES:
Research Priorities for the Nation

Edited by

William E. Larson, Leo M. Walsh,
B.A. Stewart, and Don H. Boelter

Soil Science Society of America, Inc.
Madison, Wisconsin, USA
1981

Any opinions, findings, conclusions, or recommendations expressed in this publication are those of the contributors and do not necessarily reflect the views of the sponsors.

Library of Congress Data
Soil and Water Resources: Research Priorities for the Nation
ISBN 0-89118-768-5
Library of Congress Card No. 81-52046
8107 810601

Published by the Workshop on Soil and Water Resources: Research Priorities for the Nation in the United States of America by:

Soil Science Society of America, Inc.
677 South Segoe Road
Madison, Wisconsin 53711

Printed and bound in the United States of America

CONTENTS

(Chairman and Co-Chairmen Only Listed)

PREFACE

This publication represents the best collective judgment on research priorities by a group of informed and dedicated people concerned with the future role of our soil and water resources in meeting human needs. More than 100 scientists and leaders were invited to a National Workshop held February 23–27, 1981 in Madison, Wisconsin. The goal was to identify national research priorities regarding the uses, management, and conservation of our soil and water resources. Nine working panels were established—Water, Land, Humid Regions, Semi-Arid and Arid Regions, Policy, Forest Land, Rangeland, Urban and Recreational Lands, and Drastically Disturbed Lands. A unique feature of this Workshop was the establishment of cross-review panels—Resource Base, Soil Productivity, Socio-economic Impacts, Water Quality/Quantity, and Energy Constraints—to provide interactive dialogue among panels and to make certain that these important issues were addressed. Panel chairmen prepared working papers for use at the Workshop. The papers formed the basis for discussion and provided a background for selecting research priorities. Each panel revised the working papers during the Workshop to incorporate the collective judgments of panel members, and of the cross-review panels. Panels also summarized their deliberations and listed priority research needs. The summaries provided the foundation for selecting the six most important research priorities for the nation.

These Workshop Proceedings contain an introduction, a listing of the most important national research needs, working papers of the panels, presentations of the five featured speakers, and the names and affiliations of the participants. The Proceedings are provided for those interested in an in-depth look at the material used by the panels to develop priority research needs. An Executive Summary was developed as a separate publication to present the highlights of the Workshop. The Executive Summary contains an introduction, a listing of the most important national research needs, summaries of the nine panel reports and recommendations, and a list of the Workshop participants.

We are confident that the information provided from this Workshop will catalyze the efforts of those concerned with the care of our soil and water resources.

SOIL AND WATER RESEARCH IN TRANSITION

Can we sustain the productivity of America's soil and water resources into the next century?

This question preoccupies the minds of many scientists and agricultural leaders today. It should be of concern to all Americans.

Most Americans take soil and water resources for granted. Near-phenomenal advances in agricultural science and technology have produced a seemingly unending supply of food and fiber for domestic and export uses. First there was hybrid corn, and then improved management of crops and animals, increased use of agricultural chemicals, and use of more efficient machines. Not unrelated to agriculture's enormous growth in productivity were cheap energy and favorable weather.

But the realities of American food and fiber production changed in the 1970's. What had been an era of abundance in the 1960's suddenly took on the appearance of an era of scarcity. Export demand for food and fiber crops rose dramatically, energy almost overnight became less accessible and much more expensive, and weather seemed to be more variable.

America's farmers and ranchers responded by putting an additional 54 million acres of idle land into production. The price paid by American consumers and overseas purchasers of food and fiber remained reasonable, but the toll on land and water resources proved high. National resource inventories conducted by the U.S. Department of Agriculture showed that soil erosion was excessive for maintaining productivity on a fourth of the nation's 413 million acres of cropland. In the heart of the Corn Belt, average annual loss was put at two bushels of soil for each bushel of corn produced. Increased production also resulted in the use of irrigation water at a rate that cannot be sustained in many areas west of the Mississippi River, and water quality declined measurably in many parts of rural America. Today, agricultural researchers realize to what extent advances in science and technology had actually masked the damage to the land. Soil and water resources took on the quality of finiteness.

This perception was further enhanced in the report of the National Agricultural Lands Study released in January, 1981. Competition for land intensified in the past decade, the report said, to the point that only 127 million acres of additional cropland with high or medium potential remain available. Noting that 3 million acres of agricultural land are now lost to nonagricultural uses each year, the study concluded that domestic and export needs for food and fiber will likely require that all available acres be in production by the year 2000. This study group also noted that pressure on the land will be even greater if erosion is allowed to continue to reduce soil productivity.

While much concern rightly exists about the management of our nation's cropland, similar concerns are being expressed about our range, forest, recreational, urban, and disturbed lands. The U.S. Department of Agriculture estimates that one-third of our rangeland is inadequately managed for their long-term protection. Yet we may have to increase pressure on these lands for our red-meat production as we increase pressure on cropland for our grain production. Forest lands are being damaged by erosion, soil compaction, and nutrient removal. Forest lands also are being lost to agricultural uses and mining. Yet they are increasingly being used for recreation, wildlife, and aquatic habitat, as well as for timber and paper production. As we intensify mining for coal and other minerals, the need to reclaim these lands and protect associated water resources will be great.

America's dedicated farmers, ranchers, and foresters have expressed their concern about the degradation of soil and water resources. Others are equally disturbed. Those interested in wildlife, wetlands, and wilderness have expressed their anxieties in many ways.

It was in this setting that more than 100 of the nation's best scientists and leaders in agriculture and natural resources met the week of February 23, 1981 in Madison, Wisconsin, to discuss soil and water research priorities. Knowing that public funds for research are limited, knowing that enhanced productivity is the watchword of the day throughout the nation, and knowing that most Americans think of agriculture almost solely in terms of their food supply and its cost, the charge given to participants was twofold. First, what research is needed to ensure the care and maintenance of the nation's soil and water resources into the next century? Second, how will this research make a difference in the lives of people?

Research, of course, is an integral part of efficient food and fiber production. It is inextricably linked with education, extension, and action programs. Fortunately, America's farmers, ranchers, and foresters show a remarkable penchant for innovation and using new technology.

Past efforts have concentrated on production-oriented research, and we can be proud of these achievements. This research reduces food costs at home, provides for exports to help with our balance of payments, and gives the world a few more years to bring food needs and production into balance. Furthermore, the payoff has been substantial—on the order of a 50% annual return on research expenditure.

In some respects, however, agricultural research has been a victim of its success. Society's preoccupation with the technological aspects of agriculture, coupled with the crop surpluses of the 1960's, tended to dampen the public's enthusiasm for funding research—and these factors led to an insensitivity to the exploitation of soil and water resources. Federal support for soil and water research has remained nearly constant since 1966. Agricultural productivity, which grew at an annual rate of 2.1 percent between 1939 and 1965, has since fallen to 1.7 percent. For some crops, officials in the U.S. Department of Agriculture contend that productivity has reached a plateau.

Production-oriented research must and will continue in both the public and private sectors, particularly the latter, where the short-term payoff sells well. In the case of soil and water protection, however, public research must do what private research cannot undertake because of the uncertainty in payoff. But it is precisely this type of research that no nation, including the United States, can ignore. Prolonged strength in the national economy can not rely on a public policy of resource exhaustion.

Landowners and managers have not always used the best soil and water conservation practices. This fact is understandable because such practices have not always been cost effective. Just as the farmer has been reluctant to use conservation practices because of economic constraints, the public has been reluctant to fund conservation research because of low cost-benefit ratios. However, with increasing costs of energy and additional pressures on our soil and water resources, conservation practices are rapidly becoming production inputs. This change is already evident in many irrigated areas where water conservation is reducing energy costs and also allowing more land to be irrigated. As conservation practices become increasingly important as production inputs, public support for such research should improve, and farmers will more likely use the findings.

We must readdress the role of agricultural research in relation to long-term national needs, particularly research concerned with the care and maintenance of the soil and water resources that underpin much of our agricultural productivity. To avoid doing so ultimately may prompt embarrassing questions about the consequences to farmers, to consumers, to the national economy, and to those people around the world who have come to depend upon this country for aid and a dependable source of reasonably priced food and fiber products.

The research priorities established at this workshop were based on the assumption that innovative, basic research must be continued and expanded. Unless we make a commitment to basic research and restock our storehouse of fundamental knowledge, meeting the soil and water research imperatives identified at this workshop will be impossible.

PRIORITIES FOR THE NATION

Each of the nine panel reports was carefully considered in developing the six most important research priorities for the nation. These priorities, which address the most important concerns of each panel, are as follows:

SUSTAINING SOIL PRODUCTIVITY

One-third of the cropland and much of the range and forest lands of the nation are eroding rapidly. Research is needed to quantify the relationship between plant growth and those soil attributes affected by erosion. Once we understand this relationship, we can evaluate periodically the nation's continuing ability to produce food and fiber. This, combined with a knowledge of erosion processes, will enable us to develop the best management practices for controlling erosion on specific soils.

DEVELOPING CONSERVATION TECHNOLOGY

As our soil and water resource base becomes more fully utilized, the development of conservation technology that provides adequate protection from degradation is imperative. Resulting practices must be cost-effective and energy efficient, and must protect against all forms of soil damage, including erosion, salinity, and compaction. Technology that should receive emphasis includes conservation tillage for cultivated land, management of crop residues and organic wastes, reclamation of erosion-depleted soils and disturbed lands, nutrient management of all lands, and rehabilitation and management of rangeland and forest lands.

MANAGING WATER IN STRESSED ENVIRONMENTS

Large acreages of crops are now grown under water-stressed conditions on non-irrigated land and under limited irrigation in semi-arid regions. In the future, this acreage of water-stressed crops will increase because of increasing competition from other water users, declining ground water levels, and bringing fragile lands that have low water-holding capacity into production. Consequently, we must improve strategies and practices to increase total water yield and efficient water use.

Examples of such strategies for irrigated agriculture are: increasing water-application efficiency; predicting accurately when and how much water is needed; using irrigation return flow; and using water of impaired quality. For non-irrigated areas, examples are: developing varieties and species that require less water; improving fallow and overwinter water-storage efficiencies; and developing water-conserving tillage practices.

PROTECTING WATER QUALITY

All land uses affect the quality of surface and ground water. Research must determine the relations involved in transport of sediments and chemicals by surface and subsurface flow to permit prediction and development of land management practices. The consequences of certain contaminants in water are difficult to delineate because they depend greatly on time, location, and intended use. The effects of specific pollutant levels must be assessed in terms of subsequent water use, so that arbitrary and capricious standards are not established.

IMPROVING AND IMPLEMENTING CONSERVATION POLICY

The basic factors that influence the effectiveness of soil and water conservation policies involve questions of institutional arrangements and the assignment of responsibilities among various levels of government—federal, state, and local. These arrangements influence adoption of resource conservation practices through economic and other benefits, perceived or real. But the interactions among the policies of the various levels of government on adoption of resource conservation are not clear. Research is needed to quantify the effect of alternative policies on adoption of soil and water conservation practices.

ASSESSING SOIL AND WATER RESOURCES

The nation is becoming increasingly concerned about the amount and quality of its soil and water resource base. But methods for assessing the status of and changes in the resource base are inadequate. Land and water resource data need to be consistently and accurately collected, assembled, and maintained for all parts of our nation. Technology is needed for data collection through a combination of ground, aircraft, and satellite acquisition systems, along with efficient data analysis.

Epilogue

This workshop marks a new era in cooperative planning for soil and water research. The enthusiasm and dedication of the participants and sponsoring organizations were outstanding. A program for improved communication must be initiated with scientific, producer, industry, and policy groups. Dialogue with these groups will provide background for periodic updating of research needs. Our goal is to provide the best possible research information regarding the use, management, and conservation of soil and water resources.

1

Jan Van Schilfgaarde and George J. Kriz,
Co-Chairmen

Nicolaas Bouwes, Sr., Jerry Bucheim,
James M. Davidson, Edward A. Hiler,
Arthur G. Hornsby, Howard P. Johnson,
Arnold Miller, R. Wayne Skaggs, J. Herbert Snyder,
Peter M. Tidd, Grant W. Thomas,
Martinus T. Van Genuchten

WATER
A Basic Resource

Water, a vitally important natural resource, impacts all of man's activities; not only agriculture is dependent on water, but so are mineral extraction and processing, energy generation, manufacturing, commerce, navigation, recreation, and domestic water use. Water is a renewable resource, but some is obtained from non-renewable sources. Water is easily polluted and frequently wasted.

On the average, direct daily per capita water use in the USA is about 90 gallons; only 2 of these are used for drinking and cooking, the rest for bathing, flushing wastes or watering the lawn. Indirectly, our water use is much higher; including the water used to grow our food and manufacture the products we use, the total per capita water use is about 2,000 gallons per day.

The average supply, in the form of precipitation, greatly exceeds this use. Unfortunately, water is unevenly distributed in space as well as time. The West, with 60 percent of the land, only receives 25 percent of the precipitation; the average annual precipitation in Louisiana is 55 inches, in Nevada 9 inches. From year to year, fluctuations in precipitation of 20 percent about the mean are common; fluctuations in stream flow are much larger.

Average daily withdrawals of fresh water are about 54 percent of dependable supplies (1.9 million acre-feet). Withdrawals east of the 100th Meridian though Dodge City,Kansas (520 thousand acre-feet) are roughly equal to those to the west (510 thousand acre-feet); 88 percent is returned for reuse in the East, but only 48 percent in the West.

Thus averages don't tell the story. In fact, surface water shortages have been identified in parts of all 21 water resource regions identified by the Water Resources Council in its assessment, and acute shortages are predicted in the near future for 16 of the 106 subregions (WRC, 1978). Ground water overdrafts aggravate the dilemma in parts of the country.

Not only in quantity, also in quality do we face serious problems. Pollutants may pose a public health hazard or affect stream biota, may simply increase the cost of water use, or reduce its value for agriculture. Thus, depending upon the intended use of the water, its quality or quantity may be the restraint. The effects of quality can be subtle, insiduous and hard to detect, as with intermittent or low level pollutants—or more obvious, as with salinity.

Flooding of agricultural land caused about 1,100 million dollars damage to crops and pasture in 1975. About 9 percent of the nation's cropland is on flood plains which tend to be the most productive of our soils. If current flood damage trends continue, the productive capacity of our agricultural sector will be diminished and the cost of producing food and fiber will increase slightly.

Elaborate statistics have been published as part of the Water Resources Council's Second National Assessment (WRC, 1978); also, the USDA prepared an appraisal as part of its response to the Soil and Water Resources Conservation Act (USDA, 1980). There seems to be little point repeating such information here.

Instead, we shall briefly and qualitatively sketch the nature of water-related problems as we see them. We shall address some questions of choice in selecting options towards resolving problems, and call attention to the interweaving of numerous phases of society in the ubiquitous water field. Finally, we shall raise some issues concerning research policy.

WATER QUANTITY

The timely availability of water is crucial to many phases of society. We tend to think of the eastern United States as water rich and of the West as water poor, but shortages can occur in any region or in any season. Much of the water that falls as precipitation is used in situ for biomass production. In fact, 75 percent of our agricultural production depends on it. Part of the precipitation collects in rivers or surface and subsurface reservoirs. Groundwater accounts for roughly one fourth of fresh water supplies.

The emphasis of this workshop is on resource use in agriculture and forestry. It is appropriate, however, to recognize explicitly that these uses occur with numerous other uses, whether in harmony or in conflict. Irrigated crops may be the greatest consumer of water, but irrigation must compete with many other, often higher value, off-stream uses such as municipal and industrial use, power plant cooling, and mining. In addition, in-stream uses for fish and wildlife, recreation, navigation, and hydropower generation must be satisfied, while development for intensive use may adversely affect wetlands, estuaries or other fragile components of the landscape.

Thus problems of agricultural water use, or decisions related thereto, cannot be considered without reference to other uses in the system; and economic considerations must be balanced with ecological and environmental concerns.

1. Humid Regions

Notwithstanding adequate average supplies in the humid areas of the United States, water shortages are common, as are cases of excess. Water availability has an effect on agriculture, while agricultural practices in turn affect water supplies. The management of water for crop production may call for irrigation or drainage, whereas concern for wildlife values calls for little change in the original ecosystem. One must realize that in much of the Midwest, often described as the world's breadbasket, there would not be a profitable agriculture without drainage.

Past research has revolutionized the fields of drainage materials and installation techniques. The theory of water flow through soils, developed extensively over the years, only recently is being put to routine use in rational drainage design. While drainage concerns water removal from the rootzone, irrigation is designed to add water. It is likely that the largest response to increased water management efforts, in terms of agricultural production, can be expected from irrigation in the Southeast; the primary problem is availability of reliable and economic water sources. More broadly, the need is for development of total water management systems that modulate the soil water regime to best fit the crop. Usually, there are a number of alternatives for meeting the crop and farming needs; some of these alternatives may cause more severe erosion, release more phosphorus or nitrogen to the environment or result in more pesticide in runoff waters than others. We need to identify and quantify those effects so that data will be available for reasonable political decisions. There is also a need to know the production penalty for less than optimum water management.

Water management also concerns erosion and sedimentation. Notwithstanding heroic efforts over many decades, serious problems persist. For example, in 1977 erosion rates exceeded 10 tons/acre on 32 percent of the land used for row crops in the Southeast, 9 percent in the Northeast and 19 percent in the Corn Belt (USDA, 1980). Part of the damage from erosion is gradually declining crop yield on the farm, part increased production costs; much of the damage is downstream–muddy water, silted-in rivers and harbors, increased floods. Current technology makes it possible to reduce erosion substantially from existing levels, although numerous problems are still unresolved. The recent trend towards minimum tillage has opened new opportunities (and created some new problems). However, a major problem in achieving effective conservation measures is the time frame in which the benefits or costs are realized. Given a number of assumptions, calculations can be made to show the effect of the planning horizon and interest rate on the value of a ton of topsoil lost; additional calculations can put an estimate on the damage from a ton of sediment received. Such calculations, found to be very sensitive to these parameters, illustrate why farmers may not be persuaded readily to apply conservation measures and why special incentive programs have a place. In qualitative terms, the question needing an answer is: In arriving at land management decisions, who represents future generations in the negotiations, and how are the interests of society at large harmonized with those of the land owner or, more likely, the tenant?

Agriculture, by definition, changes the natural vegetation on the land. These changes affect water use and, in particular, the timing of runoff. The ability to predict the effects of soil and water management practices on runoff timing and amounts, and search for new management options that may enhance runoff characteristics, are among the key issues in hydrology. They, in turn, are closely linked with water quality questions.

An area of concern, particularly in the Southeast, is the potentially serious consequence of substantial further development of groundwater supplies for irrigation or other uses. Localized problems of salt water intrusion illustrate the possibility of excessive development even in an area of generally very generous supplies.

2. Semi-arid Regions

The same general problem areas mentioned as prevalent in humid regions are also present, with modification, in semi-arid regions. The dominant problem in this climatic region, however, is soil water conservation. About one-half of the U.S. cropland is found in the Great Plains, where water is the most important factor limiting crop production.

Tremendous progress has been made through plant breeding, devising new tillage practices, and developing weed control practices. These practices have both increased absolute crop yields and reduced year-to -year variability in areas of marginal rainfall. Snow management technologies have improved, such as the use of grass barriers to trap snow in wheat fields. Notwithstanding the progress, a host of problems—and opportunities—remain requiring a continuing input from research.

Ironically, success in water conservation also can be credited with an increase in saline seeps throughout the Northern Great Plains of the United States and Canada. Use of wheat-fallow rotations with effective weed control often increases deep percolation enough to upset a delicate water and salt balance; solution to the problem will most likely be found in a change in cropping pattern, i.e., by introducing deeply rooted crops into the rotation that dry out the soil profile.

Again, misuse of a fragile resource can have far-reaching consequences. Prior extensive plantings of wheat contributed, during the drought of the 1930's, to the dust storms—120 at Dodge City, Kansas in the 1936-37 season—caused by wind crosion. Overgrazing also had its toll, as witnessed by nearly 100 million acres of brush in Texas. Effective use of water, a prime requirement for productive land use in these regions, depends on many subtle and not fully understood factors.

Where water supplies have permitted, irrigation has increased farm production and income stability. Irrigation especially has expanded where ground water was available, at the risk of overdevelopment and water mining. It is of interest, however, that the projections made in the Pick-Sloan program for surface water development in the Missouri Basin have not proven out; interest in irrigation development has lagged far behind anticipations and, to the extent it was developed, didn't compete favorably on economic terms (Wilkinson, 1981).

The dominant challenge in the semi-arid region is to further develop a land utilization package that will make full use of in situ water supplies without long-term harm to the resource itself. The tremendous rate of expansion of irrigation in Nebraska may or may not be sustained by a seemingly inexhaustible supply of ground water; the falling watertable in the High Plains of Texas makes it clear that, there, alternatives must be sought.

3. Arid Regions

Arid regions, by definition, suffer from a lack of precipitation. Agriculture in arid areas tends to depend heavily on irrigation and, thus, on the distribution and delivery of water that often is imported. Sustaining agriculture in the face of expanding water demands for urban and industrial uses requires the development of new supplies or the conservation—and more effective use—of existing ones. Van Schilfgaarde and Hoffman (1980) recently argued that the former course was unlikely to provide relief for an increasingly threatened irrigation agriculture, leaving the latter as the only choice.

Technically, Western water supplies could be augmented by interbasin transfers. Gigantic schemes have been proposed from time to time and it is readily conceded that the northern reaches of the continent contain huge quantities of water that could be exported. Neither economically nor politically are such schemes likely to be executed. Some argue that weather modification offers substantial opportunities for increasing precipitation; there are both technical and institutional reasons for questioning the feasibility of such schemes on a large scale. Icebergs could possibly be transported to arid regions, or seawater desalted. It appears prudent for irrigation agriculture to look elsewhere for its water.

Water conservation should not be seen as a panacea. Instead, it should be recognized as having a substantial potential for stretching existing water supplies. In cities, this may take the form of bricks in toilet tanks, meters on individual households or steeply escalating price schedules with increasing usage. In agriculture, the savings can be equally real, even if the public perception of scandalous water wastage in irrigation is frequently false.

Many, though by no means all, irrigation systems depend on a communal water collection, storage, and distribution system. Canals and reservoirs are subject to substantial seepage losses. These losses can be minimized by lining with various materials, ranging from concrete to plastic sheets to clay blankets. A USDI (1978) survey of 61 irrigation projects in the West indicated that seepage losses from their distribution systems could be reduced by 1.5 million acre-feet annually. In California, the Water and Power Resources Service currently is completing the lining of 50 miles of the Coachella Canal to save an estimated 135 thousand acre-feet of seepage water per year (Lopez, 1978). Clearly, opportunities for reducing seepage exist.

Not all consider seepage reduction a benefit. In fact, a serious question has been raised whether loss of habitat for certain species of wildlife, or loss of phreatophytes associated with water system improvements, should be mitigated, even if the environment to be saved was caused by poor canal construction originally. The question is not technical but political; still it has an important bearing on water resource use.

When evaluating water use efficiency in irrigation, one must specify the objective function. Obtaining maximum crop production per unit of water will often be in conflict with maximum economic return. Water wasted on one farm may be the supply for the next. Inducing increased seepage and deep percolation upstream may modulate the timing of river flows downstream, thus better matching water supply with water need.

Effective irrigation management requires application of the proper amount of water to the land at the appropriate time, which in turn requires a well adapted physical plant together with competent management. A combination of improved understanding and better technological hardware recently has resulted in drastic changes in irrigation practices. It has been demonstrated that more effective on-farm water use often is indeed possible and, in many cases, practical, compared to past experience. It has also been verified that such changes do not always conserve water.

Recognizing the limitations and caveats just indicated, there remain many significant opportunities for more effective use of existing water supplies through innovative application of existing technology, or through the development of new ideas. A combination of precision leveling, lateral improvements, use of metering devices, automation, and other techniques has recently increased on-farm water use efficiency some 20 percentage points on 6,000 acres in Arizona. An experiment in California showed that increasing planting density of cotton reduced evaporation losses by earlier canopy development, and reduced transpiration by earlier maturity while maintaining yield. In another test, keeping equipment off the land substantially increased alfalfa yield; presumably water use was hardly affected. Plant breeders have demonstated their skills in providing many desirable traits, be they increased yields or insect resistance; they have not put much effort into breeding for water use efficiency or, possibly even more important, salt tolerance.

Significant potential for stretching water supplies exists in substitution of relatively brackish water for fresh water. To illustrate, cotton can be grown with water containing up to 6,000 mg/liter total dissolved solids as the sole source. Such use would free water now used for cotton to irrigate salt sensitive crops. The substitution also would reduce the amount of drainage water needing disposal, an important consideration in some regions. Alternatively, such use would open up substantial stores of groundwater now considered too saline for agricultural use. Bahr (1979) estimated that, in New Mexico, an amount equal to 500 times current total annual use could be "discovered" in this manner.

Another area deserving far more attention than it has received is more intensive use of natural precipitation through water harvesting. Hardly a new concept, changes in economic forces warrant reevaluation of its potential. Near Page, Arizona, grapes have been grown successfully in terrace channels collecting runoff from the local 14 inch precipitation. In Arizona alone, there are about 2.5 million acres of land that could be used for more intensive production using similar concepts. Water harvesting also can be used to supply water for livestock, enabling more effective use of the range.

4. Energy and Water Supply

Most irrigation water is pumped. It may be pumped from a ground water reservoir. Pumps may also be used to move it through canal systems or pipe lines, or to pressurize sprinkler systems. Pumps may be required again to remove drainage water.

As pumping lifts increase and unit energy prices increase, the costs become a major factor in encouraging water conservation. In some cases, these costs even have become prohibitive, forcing abandonment of irrigation altogether.

There are other potential effects of changes in the energy economy on water supplies. The use of crop residues as a source of energy, so far contemplated but not implemented on a significant scale, could have important effects on infiltration and runoff. Similarly, use of forests for energy production from biomass could adversely affect hydrology. These possible effects must be taken into account before we embark on biomass conversion to energy with excessive enthusiasm.

It also has been proposed that crops be grown specifically as energy sources, such as for alcohol. Aside from questions of energy efficiency per se, it should be explicitly recognized that such use of crop land is not in addition to, but as a substitution for, food and fiber crops. As such it should be viewed as an alternative, competitive use of water.

The popular press—as well as a number of scientists—have proposed that desert plants can be grown on currently unused land as a source of oil (euphorbia), rubber (guayule) or as a source of energy through biomass conversion. Unfortunately, it is not always recognized that biomass production in the desert, as elsewhere, requires water. For all practical purposes, biomass production is proportional to water use. Thus, again, use of water for such purposes must be seen as an alternative use of water.

In contrast, brackish water can be used to grow halophytes, and the biomass obtained in this fashion can be used to harvest energy. In this way, otherwise idle water sources are put to constructive use.

WATER QUALITY

Although a clearcut separation of water problems by quality and quantity is not feasible, separate discussion serves to highlight the importance of both attributes.

1. Sediment

Sediment, the product of erosion, is the greatest polluter of all; the mass of sediment far exceeds that of any other pollutant. Sediment production varies widely from site to site, depending on soil type, geology, topography, land use and climate, as well as on conservation practices.

Erosion from cultivated fields is a high contributor to sediment loads in streams. Forest cover tends to offer good protection, but erosion from logging roads, from forest lands after cutting, or immediately after forest fires, can be severe. Highway construction, poorly maintained roadbanks, urban construction, and mining operations all can lead to heavy sediment production.

Sediment is entrained in and transported by moving water, deposited by slowing water. Sediment deposits were the source of the fertile lands in the Nile Delta; on a smaller scale, such deposits behind grass backslope terraces in the corn-farmed loess hills of Iowa enhance the value of the land. Reducing sediments in water may increase stream bank erosion or channel degradation. Yet the primary effects of sediment are adverse. Reservoirs fill, harbors silt up, fish yield is reduced, industrial maintenance costs are increased, and sprinkler irrigation systems fail to work. Besides these "engineering" consequences, water quality is affected by the pesticides the sediment carries, the phosphorus adsorbed on it, and salts that may be dissolved from it.

Sediment and erosion problems are not new; neither are they confined to agriculture and forestry. Substantial research investments have been made over the years that have led to increased understanding and improved management practices. They still deserve priority attention for further research. One need derives from the emphasis placed on erosion control in 208 plans and similar regulatory actions. There is a tendency to require quantitative estimates of the status quo, or of the effects of management alternatives; such quantitative estimates—although frequently made with aplomb—are often subject to very large uncertainties, or even technically indefensible. There is a valid need for the development of more and better alternative management options, but there is an equal need to better predictive capability and, especially, to enhance the understanding of uncertainty in such predictions.

2. Salinity

Of all water diverted, 90 percent is used in irrigation. In arid and semi-arid regions, irrigation is always associated with salinity. As the crops transpire essentially pure water, the dissolved salts derived from the environment remain behind, causing a gradual salination of soils and waters. Salinity is not restricted to irrigated lands, however. Seepage and evaporation have led to salt problems in many non-irrigated areas as well.

For irrigation agriculture to survive, accumulating salts must be leached out of the soil and transported down the river system, or else disposed of in areas set aside as salt dumps.

One can illustrate a range of salt-related problems by briefly considering the Colorado River. Fed by mountain streams, the water of the Upper Colorado contains little salt, say 200 mg/liter. As water progresses downstream, three processes tend to increase its salt content: concentration of dissolved salts by evaporation, displacement of stored salts by percolating waters, and dissolution of salts from soil and subsoil minerals. Use and reuse of water for irrigation contributes substantially to these processes; 37 percent of the salt load in the lower Colorado is attributed to irrigation.

The water of the Colorado is allocated by court decree and treaty to the seven basin states and Mexico. The total allocations exceed the mean annual flow. Intensive development of the river has led to increases in salinity over time and future increases are anticipated with further development, particularly that of an oil shale industry. These salinity increases lead to substantial disbenefits; the greatest costs are borne by urban and industrial users (mainly in Los Angeles), not by agriculture.

A number of options are open to reduce the salt input to the river, including changes in irrigation practices. To date, such irrigation water management schemes have proven to be far more cost effective in reducing salinity than other options considered. Water availability is obviously crucial to irrigation agriculture, but the benefits obtained—the price the farmer is able to pay—tend to be substantially lower than the benefits from urban or industrial use. Irrigation users must make every effort to make the best possible use of their water if they are to continue to compete in the (subsidized) market place.

Recent research indicates that less water is needed for leaching salts than was earlier believed necessary. Reductions in the leaching requirement permit application of less irrigation water. Progress in understanding the physico-chemical processes involved has led to the ability to better predict the effects of changes in irrigation management on the salinity status of the soil as well as of the ground water and river systems. The specific circumstances where more precise irrigation management can reduce the undesirable effects of irrigation have been demonstrated.

As we drive the system in the direction of reduced leaching and increased irrigation water use efficiency, we also reduce the allowable margin of error and increase the risk of reducing crop yield and harming the soil. Thus further research—and transfer of technology—is needed to provide better and more reliable irrigation systems, as well as to better understand, monitor, and manage the salinity processes. Furthermore, better delineation of crop tolerance to salinity and increased efforts to enhance salt tolerance genetically also are called for.

Good statistics are not available. The estimate, often made, that 25 percent of the irrigated land suffers from salinity to some degree may well be too conservative. The problem area takes on another dimension when one realizes the strong interplay between agricultural and other types of water use. Already mentioned is the economic cost of increasing salinity to urban and industrial users. Implied, if not mentioned, was an issue of international comity. Water quality in the Colorado and Rio Grande River affects both the USA and Mexico; in the Souris and Red Rivers,Canada as well as the USA. Add the increasing pressures for water supplies in energy development schemes—extracting oil from shale in Colorado or transporting coal slurries from Wyoming to the Southeast—and the problem grows. Furthermore, environmental and recreational values are also affected by water quality changes. Colorado and Wyoming, cited above, are cases in point. Here tourism and recreation depend on rivers that head in national parks (e.g., Grand Teton, Rocky Mountain, Yellowstone).

Earlier, reference was made to expanding water supplies by using brackish water to irrigate salt tolerant crops. We return to the topic here in a somewhat different context. The potential needs to be investigated for collecting high salt water—irrigation return flows, salt spring discharges, etc. and transporting these waters by separate conveyance to places of use. Recent developments make it possible to use relatively saline waters as cooling waters in power plants; other industrial uses are possible. One should not overlook, however, the possibility of using some of these waters for biomass production. At present, the upper limit for growing commercial crops at acceptable yields is about 6,000 mg/liter; through plant breeding, this will change. For now, however, halophytes abound that can be grown with far saltier water and the biomass produced in this manner could be feed stock for the production of fuels—methane or alcohol—or a source for the extraction of useful chemicals and pharmaceuticals.

3. Acid Rain

Although not a new pollutant, "acid rain" is newly recognized as a major source of water pollution. As a result of the combustion of tremendous quantities of fossil fuels, the U.S. annually discharges some 50 million tons of sulfur and nitrogen oxides into the atmosphere. Some of these oxides are converted to acids and returned to the earth's surface in rain. Actually, many other substances, including organics and heavy metals, are also deposited from the atmosphere either as dust or in water (Galloway et al., 1978).

Any number of lakes in North America and Scandinavia have become so acid that they no longer support fish life. In many others, there have been drastic changes in lake biota. Similarly, it has been estimated that forest production in Scandinavia has been decreased about 5 percent, although other studies show some doubt on the validity of these and similar estimates of damage to plant life. Because of the high buffering capacity of soil, the effect on soil properties per se is expected to be less pronounced, although changes could well come about indirectly. In short, it has been established that the adverse effects of deposition of atmospheric pollutants are drastic in some circumstances and ill defined or poorly understood in others. It is important that a more extensive data base be established to assess the extent, the nature, and the geographic distribution of the phenomenon of atmospheric deposition. We also need a better understanding of the consequences so that preventive or curative action can be devised.

4. Nutrients

In the 1960's and continuing into the 1970's, an extensive body of knowledge was developed on the fate of the major nutrients, primarily nitrogen and phosphorus, in the environment as a consequence of agricultural operations. This work was done in response to public concern with water quality issues and the need for information to write or implement new legislation. We now have a far better understanding of the chemical and biological reactions that these elements undergo, the degree to which they contribute to water pollution, and the effects of management practices thereon. We have developed mathematical models to describe the system of water flow with its constituents.

It should not be concluded that all answers are in. It seems appropriate, however, to highlight a shift in emphasis, made possible by the early answers now in hand and made attractive by a change in economic conditions. High fertilizer costs, as much as environmental concerns, have shifted emphasis to maximum plant use of available nutrients; to research on nitrification inhibitors, on increasing microbial nitrogen fixation, and on enhancing mycorrhizal associations that aid phosphorus nutrition; on plant residue management and double cropping; and on decreasing the leaching of plant nutrients by extensive irrigation. All these aspects, indirectly, will have an effect on water quality.

There also is greater interest in reuse of sewage water and land disposal of sludge, as a source of water as well as nutrients. In this area as well, the pollution potential of the major nutrients is no longer the focus of most research. The most urgent issues in water quality associated with agriculture concern the more subtle effects of minor elements and, in relation to the application of sludge, the degree of exposure to various disease vectors.

The earlier questions must not be forgotten and placed in proper perspective. Especially as regulatory programs such as 208 plans come into being, it is important to understand the strengths as well as the limitations of the models that are available to describe the present circumstance, or to predict the effect of future management options. Also, there is a need to identify and quantify the total cost and benefits of alternative management practices and cropping strategies. These total costs or benefits should be the end result of the 208 plan. For example, a proposed management practice may provide a benefit in terms of nitrate control and fish production, but a cost so far as sediment and phosphorus control, and crop production. The magnitude of the various benefits depends on the situation—and maybe time, as it could change from year to year. The ultimate regulatory decision will be political. But we must be able to draw a clear picture of the alternatives.

5. Pesticides

The pesticide situation parallels that of nutrients. Following extreme—and in retrospect, often justified—concern with indiscriminate use of pesticides, a substantial body of knowledge has developed describing the fate of various pesticides applied to the land. Models were developed to organize this understanding.

In the meantime, emphasis shifted from use of persistent chlorinated hydrocarbons applied in relatively large doses to use of more target-specific, short-lived compounds applied more judiciously, often as part of an integrated pest control program. Thus the emphasis in research has shifted also.

As long as new compounds and new pest management techniques

continue to be introduced, the need for research to evaluate their effect on water quality will stay current.

6. Water Quality and Energy Development

The effect of mining operations on water quality has long been a matter of concern. The new component in this area is the likelihood of very substantial increases in coal mining, oil shale processing, and industrial processes to extract fossil fuels; and to convert these energy sources to desirable forms and transport them.

Most of the activity will take place in areas naturally deficient in water and already subject to water quality problems. Depending on the process and location, the specific problem can vary from simply some increase in consumptive use, to ground or surface water pollution, or to disposal of low quality waters in areas far removed from their origin.

The interface with agriculture comes about in various ways. Water use for energy development is likely to be in lieu of water use in agriculture. Reestablishing vegetation after the fact, or managing spoils, will call for agricultural expertise. The knowledge base from which to predict the effect of mining operations on water quality, or to evaluate alternative ameliorative plans, is likely to be derived in part from agriculturally related fields such as soil science and agricultural engineering. These fields will continue to contribute to the solution of energy related water quality problems.

Thus, although peripheral to agricultural interests, the effect of fossil energy development on water quality cannot be ignored by those concerned with soil and water resources.

INSTITUTIONS AND WATER

Questions of water quantity and quality include many important physical and biological components, but clearly they also concern many other aspects of society. An ubiquitous—and often limited—resource that enters into almost all phases of human activity, water raises issues in economics, in law, in the social and political sciences.

Problems tend to vary regionally. In the East, emphasis is on the need for better methods to allocate water among competing uses and among adjacent jurisdictions; and for more effective coordination among institutions in planning, administering, and data gathering and processing. Heavy population pressures in metropolitan areas and, in New England, dependence on seasonal tourism with special demands for water, make explicit concern for agricultural water less important.

In the Midwest and Southeast, agriculture takes on more importance. Good water allocation decisions require planning. Planning, in turn, goes beyond water issues. It involves explicit decisions about the nature and extent of future development. Planning and allocation decisions depend on value judgments which, in part, require an estimate of the economic values gained or foregone by certain uses. Such decisions also are influenced by water rights. Ownership of water, often subject to litigation, raises questions of Federal versus State rights and of the relative effectiveness of central planning compared to local decision making.

As one moves West, the problems aren't necessarily more complicated, but they do become more explicitly crucial and irrigated agriculture becomes central. The same problems of allocation and planning, of institutional coordination and of conflicts across jurisdictional lines prevail. One must add, however, a set of problems concerning water pricing, transfer of water rights, water subsidies, and the economic importance of irrigation. More clearly than in the East, one is faced with overlapping, if not conflicting, powers and interests between Federal, State and local governments. In this region, as also in the Midwest, effective conjunctive use of surface and groundwaters requires sound technical information and, certainly, a workable institutional framework. In California, ground water law is restricted to adjudicated cases, making State water planning difficult; in Texas the drastic overdraft of groundwater leads to the question of who should bear the burden of adjustment, either back to dryland farming or to a water import scheme.

Should irrigated agriculture be maintained? Should irrigation water be priced at marginal cost, or even at cost? Can irrigation water problems be resolved in a free economic market? If the primary beneficiary of reduced river water salinity is the downstream municipal user, should irrigators bear the burden of ameliorative actions? As stated, these questions are overly simplistic and don't deserve an answer; behind them, however, lurk highly sensitive issues that need to be addressed, if not resolved. These conflicts are well illustrated for the Colorado River Basin in a book edited by Peterson and Crawford (1978); touching on engineering, economics, recreation, botany, political science and several other disciplines, the book provides insights but no solutions.

The questions that recur, in the West with reference to irrigation or in the East with reference to erosion control, are: Who is the beneficiary? Who is the loser? How can various interests be fairly represented, including those well into the future? Related is the problem of establishing values and the question whether it is indeed possible to model the system quantitatively in a meaningful way. The study by Wilkinson cited earlier demonstrates that so far we haven't, and suggests that we never will, develop a fully rational procedure of evaluation. Value judgments must always be qualitative, but they should be based on a sound basis of fact.

RESEARCH POLICY

It required a breakthrough in basic physics to ultimately permit precision grading with laser control and subsequent water savings in irrigation. Research is a continuum from basic research via applied research to technology transfer. Although no two people will agree on the precise meaning of "basic" or "applied", the concept of a continuum will not be contested. Evenson et al. (1979) demonstrated that, historically, the high rate of return on agricultural research investment can be attributed to the close association of basic and applied research workers, extension workers, and farmers. Basic research (or extension work) in isolation doesn't fare too well.

Research in the water field also must range from one end of the spectrum to the other. We called for research in plant breeding to develop more salt tolerant crops; we didn't mention basic research in plant physiology to pinpoint the mechanism involved in such tolerance. We called for greater effort to reduce erosion; we didn't mention the need for basic research in physical chemistry to better understand the nature of soil structure. Examples abound.

Balance is called for. In recent years, there has been a major pitch for increased support in the basic sciences. Capitalizing on the glamour of the double helix and recombinant DNA, spokesmen for agricultural research have emphasized the opportunities to feed the world through increased research in biochemistry. Such emphasis tends towards imbalance in another way. It overlooks the fundamental importance of the natural resources on which agriculture depends and the essentially irreversible damage that can result from mismanagement.

Furthermore, excessive stress on basic research fails to recognize the potential gain from applying the knowledge already available. For various reasons, "best practices" can be difficult to identify and are not always put to use: necessary hardware may not be readily available, economic conditions may favor other options, or the word simply hasn't gotten out. Greater effort toward transfer of technology is warranted, but not at the expense of support for basic research.

One aspect of recent research is the increased role of computer models in describing physical, chemical and biological processes, as well as in addressing various economic and water policy questions. With due regard for the power of the computer, and recognizing that all sound research starts with a hypothesis, and hence a model, we recommend caution against excessive dependence on detailed calculations based on doubtful assumptions. Indiscriminate use of modeling leads to unwarranted and highly misleading conclusions that imply excessive accuracy of results.

Unfortunately, the recent trend towards reduction of support for research has led to disproportionate cuts in budgets for soil and water conservation work. Part of the explanation may be found in the upsurge of interest in so-called basic biochemistry; part no doubt in evidence that not all soil and water research was highly innovative or productive. Yet the importance of these resources, in the short run and especially in the longer term, and the problems of shortage faced here and abroad, are such that we can ill afford to ignore them.

RESEARCH PRIORITIES

In the 1960's, the Committee on Water Resources Research (COWRR) in the Office of Science and Technology made a number of efforts to identify water related research needs, and to establish priorities and estimates of budget needs. In the process, it developed several classification schemes, with the 1966 scheme achieving some degree of permanence. Other studies have been made since, for or by COWRR, or independently. The COWRR was discontinued in 1977.

In 1980, a newly established National Academy of Sciences (NAS) Water Resources Review Committee prepared a preliminary report, Water Resources Research Priorities for the FY 1982 Budget; after reviewing reports from some 50 Water Resources Research Institutes to OWRT, and additional reports from several Federal agencies, this group will prepare a more detailed recommendation in 1981.

The NAS Committee recognized the need for a classification scheme, but did not adopt that developed by COWRR. The regional report summaries to OWRT, as a whole, didn't follow either scheme. Neither did we. We recognize the value of a common language, the convenience of easy cross reference. In part, it was a matter of timing; we didn't have early enough access to the Water Resources Research Institute projections to make effective use of them. In part, it is a matter of "fit"; neither COWRR's 1966 scheme nor the NAS Committee's 1980 scheme appealed to us as appropriate to our needs.

Here we present five areas of research we wish to highlight. They are not listed in order of priority. They seem to support, rather than contradict, recommendations by others in general terms. These research needs reflect that agriculture is the largest user of water in a nation increasingly aware of existing and impending water shortages. It is likely that agriculture will benefit from large-scale development of additional water resources because of water's higher value in competing uses. On the other hand, agriculture offers the greatest potential for water conservation because of its extensive use. Consequently, it is important to predict agriculture's real water needs and to pursue aggressively significant opportunities for water conservation.

Decisionmaking on water use requires information on physical, biological, environmental, economic, legal, and political aspects of water. Research, however, has tended to focus on individual aspects without regard to the total system. Integrated research bridging disciplines and synthesizing different objectives is urgently needed.

The greatest need in advancing effective water use may not be technical, but institutional. Institutional changes are needed to provide the incentives for individuals to use technologies that conserve water resources for the common good.

Research needs identified were:

1. Water Conservation

Research to determine the effects of alternative water management practices on crop yields and to develop accurate methods for quantifying these effects needs to be intensified. Required is knowledge of yield responses to both timing and magnitude of stresses caused by salinity and by deficient or excessive soil water.

Imaginative methods are needed for increasing the effectiveness of water use in agriculture. These include improved irrigation water-distribution systems, innovative irrigation methods, and changes in plant density, tillage, and harvesting practices. Research is needed on breeding plants for earlier maturity or for salt tolerance, use of brackish or contaminated water, and water harvesting. Realistic technical and institutional criteria must be developed for increased reuse of sewage and drainage waters.

2. Water Availability

To evaluate the adequacy of our nation's water resources, it is necessary to determine the future demands for food and fiber production and processing. Correspondingly, the availability of water supplies—with respect to time, space, and quality—to meet this demand must be determined. The development of alternative forms of energy—oil shale, coal—will increase demands on existing water supplies. Escalating energy costs have already increased the cost of supplying water to agriculture.

Research is needed to determine the effect of competing uses and rising energy costs on the water resources available to agriculture. Although inventory is needed, the primary need is to define constraints resulting from competitive demands.

3. Water Quality for Beneficial Use

Agricultural practices affect the quality of surface and ground water, but the exact linkage is not understood. The consequences of certain contaminants in these waters are hard to delineate because they depend greatly on time, location, and intended water use.

Research must develop the relations involved in the transport of toxic chemicals, nutrients, pathogens, pesticides, salts, and sediment by surface and subsurface flow. A full understanding will permit prediction of the consequence of on-farm management practices on water quality. Research also needs to be accelerated on the fate of pollutants.

The effects of specific pollutant levels must be assessed in terms of subsequent water use, so that arbitrary and capricious standards are not established. Research on interactions among components of water quality, management practices, benefits, and cost is of particular importance.

4. Institutional Constraints and Facilitators

A wide gap exists between technological know-how and practical adoption of water-conserving and pollution-abatement technology. One reason is that the benefits of applying new technology do not necessarily accrue to those who pay for them. Clearly, the existing institutional arrangements and incentives are not achieving the desired results.

Research is needed to establish the costs and consequences of water conservation practices, as well as to develop innovative institutional arrangements. Research is also needed to identify the nature and character of water rights, to evaluate the effects of changing pricing schedules, and to determine the political feasibility of making changes that would encourage more effective water use.

5. Assessment Methods

Better methods are needed to assess existing research information and data bases for use in resource utilization and policy decisionmaking. Assessments of data adequacy and integrity for the proposed use are needed. New methods must be able to quantify trade-offs among water development, water consumption, energy consumption, environmental quality, recreational use, land-use changes, net farm income, and local, state, and national social goals.

REFERENCES

1. Bahr, R. G. 1979. Saline water resources and future development in New Mexico. Proc. Inter-American Conf. Salinity and Water Management Technology, Texas A and M Research Center, El Paso, Texas.
2. Evenson, R.E., Paul E. Waggoner, and V.W. Ruttan. 1979. Economic benefits from research: An example from agriculture Science 205: 1101-1107.
3. Galloway, James N., Ellis B. Cowling, Eville Gorham, and William W. McFee. 1978. A national program for assessing the problem of atmospheric deposition. A Report to Council of Environmental Quality from NC-141, National Resource Ecology Laboratory, Colorado State University, Fort Collins.
4. Lopez, M., Jr. 1978. Status Report, Colorado River Basin Salinity Control Project. U.S. Bureau of Reclamation, Washington, D.C.
5. Peterson, Dean F., and A. Berry Crawford. 1978. Values and choices in the development of the Colorado River Basin. University of Arizona Press, Tucson.
6. U.S. Department of Agriculture (USDA). 1980. Appraisal 1980, Soil and Water Resources Conservation Act, Review Draft, Washington, D.C.
7. U.S. Water Resources Council (WRC). 1978. The Nation's Water Resources, 1975-2000. Second National Water Assessment, Washington, D.C.
8. U.S. Department of Interior (USDI), Bureau of Reclamation and Bureau of Indian Affairs. 1978. Report on the Water Conservation Opportunities Study, Washington, D.C.
9. van Schilfgaarde, Jan, and G.J. Hoffman. 1980. Future sources of water. Proc. Second National Irrigation Symposium, ASAE, Lincoln, Neb.
10. Wilkinson, John M. 1981. River-basin plans and realities: A Post-audit of the Pick-Sloan Missouri Basin Project. Proc. XIth Congress, Int. Congr. Irrigation and Drainage, Grenoble, France.

2

Klaus W. Flach and Chris Johannsen,
Co-Chairmen

Richard W. Arnold, Kurt W. Bauer,
Michael F. Brewer, D. Keith Cassel,
Laurence R. Jahn, Dennis R. Keeney,
Robert H. Miller, D. David Moyer,
Gary W. Petersen, Richard E. Witmer

LAND
Resource Base and Inventory

Dramatic changes have taken place in American farming during the last few decades. Our nation is moving from an era of abundance of farm products to one of scarcity. Americans are becoming concerned about our continuing ability to produce enough food for domestic and foreign markets and about the impact of modern farming on the quality of the environment.

To decide whether public funds should be spent to combat erosion, information is needed on the effect of soil erosion on food and fiber production and the social cost of resultant pollution. Better methods are needed to monitor changes in soil qualities, land use, and management; to quantify the effect of such changes on productivity; to assess the impact of changing economic conditions on the resource base; to document the Nation's ultimate capability to produce food; and to identify alternatives of agricultural production. These assessments must be accurate, relevant, cost effective, and in a suitable format for public and private decision makers. This document identifies research needed to achieve these objectives.

THE STATE OF THE LAND RESOURCE BASE OF THE UNITED STATES

The United States has finite land resources. What will happen to this basic resource as our population grows and farm exports increase? Will we have enough suitable land to grow our crops, graze our livestock, and grow our trees? Are we in danger of losing crop production to soil erosion and falling water tables? How much good farmland will be lost to urbanization and highway construction? Can we afford to lose more wetlands?

These are some of the questions that have been addressed by the U.S. Department of Agriculture during the past 2 years as a result of the Soil and Water Resources Conservation Act of 1977. The responses to these questions have provided the groundwork for some new ideas about the kind of conservation programs that Americans will need to meet future demands on soil resources. This paper presents some background information on our land resources, identifies current problems and constraints, and suggests where research is needed to answer these questions. Unless otherwise stated, the data source is the SCS National Resource Inventory conducted in 1977 (USDA, 1979).

A. Land Quantity

1. Land Use

In 1977, the 752 million acres of federal land in the United States and the Caribbean Area (Puerto Rico and the Virgin Islands) were 38 percent forest, 58 percent range, and 4 percent other land (Forest Service, 1980). In the same year, the 1,513 million acres of nonfederal land and small bodies of water in the United States and Carribbean area was 27 percent cropland, 9 percent pastureland or native pasture, 27 percent rangeland, 25 percent forest land, and 12 percent other lands, including urban areas, roads and highways, areas of open water of less than 40 acres, and streams less than one-eighth mile wide.

a. Forest Land — Forest land provides wood, water, wildlife, herbage, and recreation. Its productive capability depends on the soil, climate, topography, and associated vegetative cover. Differences in the physical characteristics of soils, the supplies of moisture and nutrients, topography, and management all contribute to the diversity of this resource and the benefits that we receive.

Forest land is defined as land under at least 25 percent tree canopy, or land that is at least 10 percent stocked with forest trees of any size, including land formerly forested that will be naturally or artificially reforested. It does not include transition zones between heavily forested and nonforested land that can also be defined as rangeland, nor does it include forested areas that can be defined as urban or built-up land. As a result of these exclusions, the acreage of nonfederal forest land shown in the 1977 SCS National Resource Inventory is smaller than that shown in some other inventory reports. The area classified as forest land according to the above definition occupies 663 million acres of the 2,265 million acres of land in the Nation.

There are 286 million acres of federal forest land and 377 million acres of nonfederal forest land. Nonfederal forest land is land that is mainly under private ownership, but it also includes state and municipal land. States in the eastern half of the Nation contain nearly 83 percent of the nonfederal forest land, and the western states contain 17 percent. The South Central Region in the East contributes 26 percent of the total, the Southeast Region 20 percent, the Northeast Region 19 percent, and the North Central Region 18 percent. The Pacific Northwest—Alaska, Washington, and Oregon—contribute 8 percent, the Rocky Mountains 5.5 percent, and the Pacific Southwest 3 percent. The Great Plains region has very little nonfederal forest land, only one-half of 1 percent of the national total.

b. Rangeland and Pastureland — Rangeland is defined as land on which the natural potential (climax) plant cover is principally native grasses, forbs, and shrubs valuable for forage. It includes grasslands, savannas, and certain shrub and forb lands. Except for brush control, it is managed primarily by regulating the intensity of grazing and the season of use. If the land is revegetated, the improved forage cover is managed like native vegetation. Generally it is not fertilized, cultivated, drained, irrigated, or mechanically harvested.

Pastureland is used primarily for the production of domesticated forage plants for livestock. The forage is generally a perennial grass or a mixture of one perennial grass and one or more legumes. In 1977, there were 134 million acres of pastureland in the United States and the Carribbean area. This compares with 102 million acres in 1967 (USDA, 1971). In 1977, there were 5 million acres of irrigated pastureland. This was 3.7 percent of the total amount of pastureland.

There are 414 million acres of nonfederal rangeland in the United States and the Caribbean area. Approximately 99 percent of this rangeland is in Alaska and the 17 western states. The other 1 percent is distributed throughout the Minnesota to Louisiana tier of states, in Florida and the Caribbean area, and as marsh range across the Gulf Coast States. In 1977, the latest year for which data are available, 68 percent of this land was well managed and 32 percent was overused. Generally, overuse reduces forage production and increases susceptibility of the soil to erosion. U.S. Department of Agriculture estimates indicate that range conditions have improved considerably in recent years. Between 1963 and 1977, the percentage of range in excellent condition increased from 5 to 12 percent, the percentage in good condition increased from 15 to 28 percent, the percentage in fair condition increased from 40 to 42 percent, and the percentage in poor condition decreased from 40 to 18 percent. In 1977, range conditions were improving in 13 of the 17 major range states, were static in 3, and were declining in 1. In spite of recent improvement, however, 60 percent of the nonfederal rangeland is still in poor or fair condition.

About 22 million acres, or 5.3 percent, of the Nation's rangelands are classified as prime farmland, of which 87 percent is in Texas, Oklahoma, Kansas, and Nebraska. This is among the most productive grassland in the United States. In addition, almost 40 million acres, or 39 percent of the Nation's pastureland and native pasture are prime farmland. It is highly probable that a large part of these 62 million acres of rangeland, pastureland, and native pasture will eventually be used as cropland. Assuming that rangeland produces 0.8 animal-unit-month per acre, converting 22 million acres of rangeland to cropland would cause the loss of 17.6 million animal-unit-months of forage. Assuming that pastureland and native pasture produce 3 animal-unit-months per acre, converting 40 million acres of pastureland and native pasture to other uses would cause the loss of an additional 120 million animal-unit-months of forage.

c. Cropland — In 1977, there were 413 million acres of cropland in the United States and Carribbean Area. This includes land in row crops such as corn and cotton, close-growing crops like wheat, orchards, vineyards, legumes and grasses harvested for hay, and summer fallow.

The National Resource Inventory showed that of the 105 million acres of wet soils used for cropland in the United States, about 34.2 million acres have inadequate drainage systems for maximum crop production (USDA, 1979). Crop yields on well drained soils in many places are 50 to 100 percent above yields on similar soils that are poorly drained.

2. Land Use Trends

Use of the Nation's nonfederal land is changing. The acreage of cropland and forest land is decreasing. The acreage of pastureland, native pasture, and rangeland is increasing. The acreage of urban, built-up, and transportation land is increasing rapidly.

The acreage of nonfederal land used for cropland in the United States has decreased steadily over the past 20 years (USDA,1971). In the period 1958 to 1967, cropland decreased from 447 million acres to 431 million acres. In the period 1967 to 1977, there was an additional 18 million-acre decrease in cropland.

The acreage of irrigated cropland has greatly increased. In 1958, 37 million acres were irrigated (USDA, 1961). By 1967, 44 million acres were irrigated (USDA, 1971) and, by 1977, 56 million acres of cropland were irrigated (USDA, 1979). The rate of cropland brought under irrigation accelerated from 0.8 million acres per year during 1958 to 1967 to 1.2 million acres per year during 1967 to 1977.

The Potential Cropland Study (Dideriksen et al., 1977) shows that the land in crops today is not necessarily the same land that was cropped in previous years. Between 1967 and 1975, 53 million acres shifted from cropland to pasture and range and 32 million acres shifted from pasture and range to cropland. During the same period, 11 million acres shifted from forest land to cropland and 8 million acres shifted from cropland to forest land. Approximately 13 million acres shifted from cropland to other farm and nonfarm uses and nearly 6 million acres shifted from other farm and nonfarm uses to cropland. A total of 5 million acres shifted from cropland to water or urban uses in that 8-year period.

Land is being converted to use for residences, factories, shopping centers, and roads at an increasing rate. During the period 1958 to 1967, about 17 million acres of rural land were converted to urban, transportation, and water uses (U.S. Bureau of the Census, 1970; Dideriksen et al., 1977). This amounted to 1.9 million acres per year. During the 1967 to 1975 period this rate increased to 2.9 million acres per year.

3. Land Distribution

Not all parts of the Nation are equally blessed with high quality soils. The Corn Belt, Lake States, Northern Plains, and Delta States have the largest acreages of soils well suited for agricultural purposes. In the Corn Belt, 76 percent of the acreage is suitable for continuous cultivation (USDA capability classes I-II). In the Northern Plains, Lake States, and Delta States, more than 50 percent of the acreage is suitable for continuous cultivation. In contrast, the acreage suitable for continuous cultivation in the Mountain States is only 18 percent and 25 percent for the Pacific States.

The distribution of soils that are of marginal value as cropland (class IV) is fairly uniform, averaging 13 percent nationwide. In the Southeast and the Lake States, however, these soils represent 23 percent of the land area. Soils suitable for grassland occur throughout the Nation. Most soils in the East and a significant acreage in the Pacific and Mountain States are suitable for woodland.

The Nation has a significant—but finite—amount of prime farmland. There are about 345 million acres of land in the United States that are considered prime farmland. Prime farmland, as defined by USDA, is the land best suited for producing food, feed, forage, fiber, or oilseed crops. It has the soil quality, growing season, and current moisture supply needed to economically produce a sustained high yield of crops when it is treated and managed using acceptable farming methods. Compared to other farmland soils, prime farmland soils produce the highest yields and require fewer inputs of energy and economic resources. Furthermore, farming these soils results in the least damage to the environment. Prime farmland areas may now be in cropland, pasture, woodland, or other land, but not urban and built-up land or water areas. They must either be used for producing food or fiber or be available for these uses.

Prime farmland soils have an adequate and dependable supply of moisture from precipitation or irrigation. They also have a favorable growing season and acceptable acidity or alkalinity. They have few or no rocks and are permeable to water and air. Prime farmland soils are not excessively erodible or saturated with water for long periods and are not flooded during the growing season. The slopes range mainly from 0 to 6 percent. More detailed information about prime farmland can be found in the Federal Register, Vol. 43, No. 21.

About two-thirds of the 345 million acres of prime farmland were cropped in 1977. Of the remainder, 42 million acres were forest land, 40 million were native pasture and pastureland, 22 million were rangeland, and 11 million were used for other land uses. From 1967 to 1975, about 8 million acres of prime farmland were lost to irreversible uses (Schmude, 1977). About 6.5 million acres were converted to urban uses and 1.5 million acres to water areas.

B. Soil Moisture and Temperature

Regardless of other attributes of soils, no plant growth is possible without soil moisture. Soil moisture must be considered in selecting suitable plants. Because of rainfall patterns most soils east of the Mississippi River contain enough moisture for crop production if they have appreciable capacity to store moisture (i.e., reasonably deep, permeable, not too steep, and not too sandy). The more frequent the rainfall, the less the storage capacity that is needed.

Moisture supplies tend to be less in soils west of the Mississippi, except in the higher mountains and the areas adjacent to the Gulf of Mexico. East of the Rocky Mountains, most precipitation falls in the summer. West of the Rocky Mountains, most of the precipitation falls in the winter. For example, the Central Valley of California receives virtually no rainfall from May through October.

Soil temperatures are also important. The commonly cultivated crops grow little until the soil temperature exceeds 41 degrees F. Alaska and many areas adjacent to Canada have a mean annual subsoil temperature lower than 47 degrees. These areas have a frigid, pergelic or cryic temperature regime and a relatively short growing season, usually less than 90 days. In Alaska, except for coastal areas of the southern part and where the native vegetation and the "0" horizon has been removed in the process of opening up new areas for cultivation, mean soil temperatures are usually less than 32 degrees F and the subsoil is frozen most of the year.

In the central part of the United States, the mean annual subsoil temperatures range from 47 to 59 degrees F. This area has a mesic temperature regime and a growing season that may exceed 6 months. The southern tier of states, where the mean annual subsoil temperature ranges from 59 to 72 degrees F, has a thermic temperature regime. Southern Texas, Southern Florida, much of Hawaii, and the Caribbean Area have mean annual subsoil temperatures that exceed 72 degrees F and have a hyperthermic temperature regime and a continuous or nearly continuous growing season. All of Hawaii, except the higher elevations, and the Caribbean Area have little seasonal variation in temperature. These two warmer areas provide more opportunities for multiple cropping and for growing crops that require long growing seasons such as cotton and sugar cane. Most coastal areas have less seasonal variation in climate than the rest of the country.

AREAS NEEDING RESEARCH

A. Characterization of soil quality

Soils in the natural landscape vary in their qualities i.e., some soils are obviously more productive than others for specific crops. Subjectively, the farmer is aware when soils decrease greatly in productive capacity because of excessive soil losses by erosion, failure to maintain soil organic matter levels, or failure to adequately maintain soil fertility. Likewise specialists in agronomy, range sciences and forestry have little difficulty in recognizing gross changes in soil productivity. These same specialists do, however, have a great deal of difficulty quantifying soil qualities for specified uses. This limitation becomes a serious problem if those individuals and groups interested in maintaining or improving our soil resources cannot adequately demonstrate in quantitative terms the true effects of poor soil management or farming practices. A farmer would more likely adopt conservation practices and the public sector would have a better basis for making decisions if the true impact of practices on soil productivity and other soil qualities could be calculated. The next sub-sections deal specifically with various factors affecting soil qualities, particularly the soil quality referred to as soil productivity.

1. Estimating Soil Erosion by Water

The Universal Soil Loss Equation (USLE) is the primary tool for predicting probable soil loss (Wischmeier and Smith, 1978). The equation is as follows:

A = RKLSCP, where A = average annual soil loss, R = a rainfall erosivity factor, S = slope steepness, L = slope length, K = an erodibility value, C = a cover and management value, P = an erosion control practice value.

The estimated soil loss A is then evaluated against T, the estimated soil loss tolerance. Tolerance values have been established for all major soils in the United States. The factors C and P are outside the purview of this document; slope steepness (S), and slope length (L) are considered soil properties in the United States and R is an energy term related to rainfall intensity and duration.

The USLE was standardized against the results of long term experiments on soil erosion on cropland in the Midwestern and Southern States. To provide information on conservation needs outside this region, the use of the USLE is being extended to cropland soils in other areas and to soils in forest and range. However, the USLE has not been adequately tested for these conditions. Problems also occur because experiments used in the development of the USLE reflected a limited range of soils, plant, cover, and rainfall patterns.

Recently, the USLE has been used to predict sediment loads in streams. This adaptation requires basic conceptual changes in the K and T factors.

There is an urgent need to expand the network of actual erosion measurements to cover all parts of the country and all major management and land use conditions to which the USLE is being applied. Methods must be developed to make these measurements in an accurate and cost-effective manner.

a. Develop Better Means for Measuring Actual Soil Loss — In the classical erosion plot all sediment, or a fraction, carried by runoff is collected and weighed. Such plots require constant attendance and represent somewhat artificial situations. They are expensive to install and maintain for the many soil and management conditions for which soil loss data are needed.

There is an urgent need for new technology to measure changes in a plot of land because of erosion. Three dimensional scanning with laser equipment might be one avenue for making such measurements. Other possibilities likely exist and should be explored.

b. Re-define Allowable Soil Loss (T factor) and Soil Erodibility Values (K). — Soil loss affects the productive capacity of the soil and sediment carried from the soil affects the quality of off-site soil, water, and air resources. The two concepts need to be separated.

A proposal has been made by W. E. Larson (1981) to establish a T1 value related to the change in productivity and a T2 value related to the impact on the environment. Additional changes are necessary in the definition of the T1 value. Past and current definitions relate to the rate of soil formation that would compensate for erosion or on sustained productivity. Neither concept can be precisely defined. Rather, emphasis should be on a T1 value that reflects how a unit of soil loss decreases productivity. Emphasis should be as much on the relative sensitivity of a soil compared to other soils, as on absolute amounts lost, to enable decisionmakers to make better judgements in allocating conservation resources.

Soil erodibility (K) should be separated into two values (K1 and K2) to reflect soil loss as it affects productivity and as it affects potential suspended sediment. This would make a clear separation between fundamentally different uses of the USLE.

2. *Estimating Soil Erosion by Wind*

The wind erosion equation estimates the susceptibility of soil to wind erosion (USDA, 1961).

The equation is E = IKCVL, where E = potential annual soil loss, I = soil erodibility, K = a soil ridge roughness value, C = a climate value, V = vegetation, and L = unsheltered distance.

A major difficulty in using this equation is presented by the I value which combines intrinsic soil properties and effects of soil management practices. If the formula is to be used analogous to the USLE the I of the wind erosion equation must be separated into two factors equivalent to K and C of the USLE. There is also a need for more information on the effect of wind erosion on yields and crop performance and for methods for measuring the intensity of the individual wind erosion events.

3. *Effects of Soil Erosion on Productivity.*

Information is needed on the effects of soil erosion on soil productivity to justify the use of private and public funds for soil conservation and for establishing objective standards of tolerable soil loss.

Numerous data on how soil erosion influences productivity were generated during the 1930's but they were obtained under extensive management. In contrast to modern farming practices, little attempt was made to compensate for the loss of plant nutrients through the application of fertilizers. Erosion, however, causes changes in productivity that cannot be compensated by additional fertilizer application. Such changes are caused by decreased water infiltration, decreased water holding capacity of the soil, a less favorable root environment, and changes in soil temperature due to changes in albedo and other factors. The quantitative effects of such changes are not well known.

Field experiments should be established on a wide range of soil and climatic conditions and with several major crops. Such experiments should be carefully instrumented to make it possible to identify the specific soil factors responsible for changes in productivity. From this information the effect of erosion on the productivity of related soils can be estimated. Knowledge gained from such experiments will also improve our capability to identify factors that limit productivity on soils not affected by erosion and assist in maximizing soil productivity under a wide range of conditions.

4. *Water Holding Capacity*

The water in soils that is available for plant growth (available water capacity) is defined as that water retained in soil pores against gravity yet not so tightly held that the plants cannot remove it. Soil factors which affect the available water are soil depth, soil texture, soil structure, soil bulk density, the presence of root restricting physical barriers (such as clay pans or tillage pans), and chemical barriers (toxic levels of aluminum). Thus, as the amount of topsoil is reduced through erosion, or as compaction destroys soil structure, the potential available water content declines. This can have major effects on the yields of crops in rain-fed agriculture, or the water use efficiency of irrigation.

The major solutions involve maintaining or improving soil structure through organic residue management, crop rotations, and controlling erosion. But to recommend these management schemes requires an objective measure of the loss in water storage capacity, and also an inventory of lands in danger of major loss of production due to a lowering of available water capacity.

5. *Loss of Nutrients*

Soil degradation which results in loss of topsoil and associated nutrients in the clay and organic matter fractions, whether by erosion or by intensive cropping, is a major problem arising from today's agricultural systems. Preferential erosion of the soil colloid fraction (clay and organic matter) may result in two to four times greater concentrations of nitrogen and phosphorus in the eroded sediments than in the bulk soil. In many cases, these losses of nutrients are overcome by adding more inorganic fertilizer. Thus, the problem often goes unnoticed until soil degradation becomes serious. Restoration of soil fertility can take many years and much expense. In the meantime, a production resource has been lost.

Any land quality inventory must delineate the amount and location of these degraded soils, and describe their relative susceptibility to degradation. Marginal lands arc especially susceptible, but any soil can be degraded if mismanaged.

Numerous easily measurable chemical properties of soils, such as organic matter content, the rate and extent of phosphate sorption and release, potassium supplying capacity, acidity, buffering capacity, and cation exchange capacity, can be characterized. With these data, and with physical property data, soil productivity can be predicted under a given climatic or irrigation regime by using results of available yield-response field trials. However, we have little information on soil productivity in relation to the properties of eroded soils. Are yields actually reduced, or can input of chemical fertilizers, etc., sustain production? If yields are reduced, what are the economics of this loss of topsoil?

An important erosion control technology is conservation tillage or a similar other system that maintains crop residues on the surface. While general agreement exists that these practices reduce erosion, increase soil organic matter, and maintain or improve soil tilth, much research remains to be done to provide complete, efficient, high-yielding residue management systems. Such systems at the same time must minimize fertilizer use and overcome problems of phosphorous and potassium losses in surface runoff, acidity buildup in the surface soil from nitrogen fertilizers, and excess nitrogen loss from dentrification and leaching.

These factors need to be incorporated in the overall land resource inventory in explicit terms where possible, so that we can accurately assess lands in need of various management schemes. Failure to recognize and correct soil quality degradation will lead to major declines in soil productivity.

6. Soil Tilth

Soil tilth is a term which describes and integrates a series of physical conditions of the surface soil. Soils with good tilth are well aggregated, possess a stable soil structure, have a high infiltration rate, have a high water holding capacity, are resistant to crusting and are easily tilled. Soils with good tilth are not only more productive, they also require less energy to till than soils with poor tilth.

An important factor in improving and maintaining soil tilth is the presence of soil organic matter. The organic contents of soils vary because of differences in organic matter accumulation during soil formation as influenced by climate, parent material, topography, biota, and time; and because of soil management practices which are applied, e.g. type of tillage, cropping patterns, erosion control, residue management, and availability of organic waste materials.

While there are some conceptual ideas about soil tilth, and how it can be improved, it cannot be defined in quantitative terms. This limits our ability to place a value on organic matter for improving soil tilth as well as on the possiblility of assigning a loss value when soil tilth decreases because of poor management.

Research should be conducted on methods for quantifying soil tilth. One approach may be to describe tilth as a function of a series of soil physical measurements, e.g. draft of a tillage instrument, crusting, infiltration rate as a function of management practices, and the impact of rain and wind. Follow up studies could include a comparison of tilth measurements to soil productivity in the absence of limiting nutrients and moisture. Changes in tilth when different kinds and quantities of organic residues and wastes are incorporated in soils could also be measured.

B. Strategy for Collecting Basic Information on U.S. Soils

The agricultural communities, including their scientific bases, at universities and research centers have largely focused their research attention on crop and animal production. Because of this emphasis a technology to monitor or assess the status and changes in the well-being of our soils remains undeveloped. Yields are always contemporary issues at both the national and international levels and little attention is given to the relatively slow, subtle changes occurring across the landscape that eventually degrade the quality of resources and destroy a major economic base of our country. It may be surprising to some and disconcerting to others that we currently can't accurately estimate the rate at which water moves or chemicals leach through the tilled fields of our nation.

In the United States the transfer of agro-technology is primarily based on soils information provided by the National Cooperative Soil Survey. Soils throughout the United States are recognized, classified, mapped, and named according to standards. Experience and research results on named soils can be extrapolated to the same and similar soils on which the research and experience has not yet taken place.

Physical, chemical, mineralogical, and biological soil properties vary as a result of long term effects of soil formation and are further influenced by effects of relatively short-term soil management. For example, lateral soil variability, i.e., variability over the landscape, is partially due to changes in slope. As a result of slope differences, variation in soil properties develop. The degree of variability, however, is not identical for each property. For example, the percentage of clay may change two or three percent over a given lateral distance, while saturated hydraulic conductivity may vary ten to several hundred-fold.

Over 10,000 individual soils are found in the United States. These soils have an extremely wide range of physical, chemical, mineralogical, and biological characteristics. Information for each soil is necessary to make intelligent decisions regarding land use. It is prohibitively expensive to collect the exhaustive sets of basic information required for each soil. While some information is available for many of these soils, no strategy exists for collecting the more sophisticated basic information, such as soil erodibility, hydraulic conductivity, and engineering properties. Neither is a system available for expedient storage and retrieval of the data.

A basic strategy for collecting information on soils of the United States must be developed. New and innovative data collection techniques must be developed to allow for more adequate and more efficient data acquisition. This would include additional research on various types of remote sensing systems. More emphasis also should be placed on the use of data acquisition systems that allow for real-time satellite data collection.

Most soil parameters, such as those for measuring tilth, K values of the USLE, or yield functions that have been mentioned before are not likely ever to be measurable by remote sensing. They require expensive measurements of field experiments that can be performed at a relatively few sites. A well designed network of sites for sample collection and for field experiments must be established that will cover a cross section of key soils in the nation in such a way that equivalent values for related soils can be estimated with reasonable certainty. Such a network would also require agreement on standard methods of measurement, standard experimental designs, and a standardized set of supplemental data such as standard weather data for fertility experiments.

Data systems, sophisticated multivariate analyses, and modelling procedures must be developed concurrently to optimize procedures for projecting information from the sampling sites.

C. An Integrated System To Handle Research Data On Soils

Several federal and state agencies have authorities to inventory, assess, monitor, and assist in managing various units of the landscape. These authorities were established at different times, and each agency developed different methods for collecting data on soils and related landscape features. At present, there is no common data base for named soils in the United States and there is a critical shortage of standard soils information to complete mandated and needed management plans for the nations' forest land, rangelands, urban areas, control zones and wildlife areas, as well as agricultural lands. Further, most resource data files are not automated, are located at a variety of widely separated sites, and cannot be used in completing resource evaluations in a timely manner.

We recommend that a national, standard, integrated, automated system be established for collecting, storing, and obtaining information on soils and related landscape features. One overall objective should be to have all information on the physical, chemical, biological, and production properties of named soils; their engineering behavioral characteristics; as well as yield and other performance data placed in one data system and available through easily accessible means.

The proposed integrated data system on named U.S. soils holds substantial promise for achieving greater efficiencies and effectiveness in planning and carrying out resource development and management programs and projects.

D. Models for Reacting to Forces Outside U.S. Agriculture

In recent attempts to model global issues (U. S. Council on Environmental Quality, 1980) it was apparent that the United States does not have a coherent and comprehensive capability to ascertain the myriad of complexities that affect agricultural production.

The forces external to the United States have significant impact on policies that control the responses of American agriculture. A poor wheat harvest in the Ukraine, increase of OPEC oil prices, sugar commodity trading, and world bank loans for project development are examples of events that influence world or foreign policies that often are translated into land use responses in the United States. As the third world increases its food and fiber production, it will change the demands on American agriculture.

Similar changes of national or sector policies in the United States also are reflected in ever changing responses in agricultural production. The period of farm product surpluses was also one of weather stability and thus many of the management and other resource inputs of that time may no longer be valid for extrapolation to new situations.

The shifts in agricultural production that result from the interaction of forces external or beyond the agricultural sector are not well enough understood to predict the consequences and develop meaningful programs to stabilize and balance our resources in agriculture.

Constraints to sustainable yields are being determined by on-going research throughout the nation yet there is not a central or unified way to bring such information together and use it to project results of other policies or events that ultimately affect agricultural production in the nation. Information gathered and analyzed by most federal, state, and local agencies and institutions has significant consequences on the analysis and conclusions obtained by others.

The solution is to develop comprehensive mathematical models to analyze the effects of forces influencing the use of the Nation's soil and water resource base by integrating the necessary data sets and having them interact in meaningful ways. The models would provide alternative strategies for realistic predictions to guide political action.

It is recommended that the U.S. Government take steps to promote a comprehensive system drawing on the expertise and knowledge from all sectors. A lead agency should be designated to coordinate the research and development activities of the many contributors.

E. Integrated Land Information System

A wide variety of land resource data in the U.S. currently is being collected, analyzed, and maintained by numerous agencies at various levels of government. These data are typically collected and used for a single purpose at a given point in time. Therefore, they are usually not compatible for other needs and users, either by other agencies at a parallel level of government, or by other levels of government. The following examples illustrate this point and the type of problems that often result.

Earlier in this paper, several federal surveys and studies that produced data by broad land ownership and land use classes were discussed. These included the Reclamation and Conservation Act, Conservation Needs Inventory, Natural Resource Inventory, and the 1976 Potential Cropland Survey. Each of these efforts produced a set of data needed for a particular agency or policy analysis. However, these data are usually inconsistent both between surveys and with data collected by other federal agencies such as the Bureau of the Census. The data in Table 1 illustrate several inconsistencies in land ownership and land use.

Similar situations exist at the local and state level. To continue with land information systems such as the above has several important consequences:

1. Data collection costs are higher than necssary, due to overlap.
2. Critical data needed for detailed analysis and complex tabulations (e.g. cross tabulations) are often not available.
3. Most importantly, trend analysis cannot be carried out, due to the non-comparability of data.

The solution is to develop an integrated land information system for the United States. The integrated system should include resource data, detailing the physical, cultural, and ownership characteristics of all land. These data should be consistently and accurately inventoried and maintained for all land, public and private, of the United States.

Table 1. Illustration of inconsistencies in land ownership and land use data.

Item	USDA publication†		Difference	
	Panel 2 paper	AIB no. 422		
	million acres			%
Federal land	752	761	9	1
Non-federal	1,513	1,503	10	0.6
Cropland	413	466	53	13
Pasture & rangeland	548	439	109	25
Pasture	134			
Rangeland	414			
Forestland	377	471	94	25
Misc. & water	175	127	48	
Misc.	--	127		

† Gene Wunderlich, Facts About U.S. Landownership, U.S. Department of Agriculture, ESCS, Agricultural Information Bulletin No. 422, November, 1978.

The integrated land information system would contain a wide range of data including type of ownership, tenure, residence, land value, tax rate, ground cover, land use, and soils information. In general, all data should be related to a specific geographic location.

Much of these data are currently collected and stored for all or part of the country. New technologies should be implemented as appropriate. Development of new technologies for collection, storage, and manipulation of land data should be undertaken as necessary. For instance, remote sensing technologies should be evaluated to ensure that spectral and spatial resolution of the system is optimal for each land factor and data use. Frequency of data collection also should be evaluated for various uses and users.

To implement an integrated land data system throughout the country, three actions by the federal government are needed (National Research Council, 1980).

1. A federal agency should be designated to coordinate the systems development and operation.
2. Research and technical assistance should be sponsored by federal agencies to identify and ensure implementation of the appropriate technology.
3. Authorizaton of a cost sharing program to assist and encourage local and state governments to develop compatible land data systems.

Once implemented, the integrated land information system will have important consequences for all levels of government. At the local level, title transfer, property taxation, and other land data will be maintained in a more usable format at costs that are similar or less than those incurred for the present fragmented system. At the federal level a standardized integrated data system will provide the most cost-effective way to define more clearly particular resource issues where a policy decision is needed. It will provide the capability to quickly obtain estimates of the impacts of proposed alternative solutions to a particular policy problem. Answers can be obtained to hypothetical questions by manipulation of data stored in the system, and use of an integrated land information system should improve the quality of the policy that is finally selected for implementation.

The system will also provide the basis for explanation of the policy to the general public. A clear, concise explanation of the soil and water conservation policy to be implemented, the reasons it was selected, and government expectations of the policy should result in a clearer understanding and more rapid acceptance of the policy.

RESEARCH PRIORITIES

American agriculture has changed dramatically during the last few decades. Our nation is moving from abundance to scarcity of farm products. Americans are questioning our continuing capability to produce enough food and fiber for domestic and foreign markets and are concerned about the quality of the environment.

The National Agricultural Lands Study of 1981 estimates that between 1967 and 1977 three million acres of farmland were lost each year. If these estimates are correct then all potentially suitable farmland will be in use by the year 2030. The National Resource Inventory of 1976 showed that erosion on one-third of our cropland is greater than thought to be permissible without reducing future productivity.

Many believe that erosion, loss of organic matter, and the impact of heavy machinery are slowly deteriorating our surface soils, but we do not know how to evaluate the deterioration adequately. Crop yields do not yet reflect this soil deterioration because of the beneficial effects of fertilizers, pesticides, better crop varieties, and other technological improvements. However, compensating effects of these inputs may be smaller in the future.

Information is needed on the social costs of erosion to decide whether public funds should be spent to combat it. Social costs include pollution and loss of soil productivity. A recent evaluation of the agricultural conservation program indicated that public funds have not been used effectively in the geographic areas of greatest need.

Research must develop better methods to monitor changes in soil attributes, land use, and management; to quantify the effect of these changes on productivity; to assess the impact of changing economic conditions on the natural resource base; and to identify alternative systems for agricultural production. Unfortunately, research and assessment are expensive. More economical ways to collect and analyze data are needed.

Following are the areas of research related to the nation's land resources that need immediate attention:

1. Characterization of Soil Quality

A quality of soil, as used here, is a value judgment about the status of a soil under specified use and management. We need more precise definitions of various soil quality parameters to recommend best land uses and to predict the costs of erosion and intensive tillage. We need to know what effect the loss of topsoil has on soil productivity through loss of plant nutrients, reduced rooting depth, loss of water holding capacity, and degraded soil tilth. We must develop management strategies that are technically and economically sound to overcome the loss in soil productivity.

2. Strategy for Collecting Basic Information on U.S. Soils

We have some 10,000 different kinds of soils in the United States for which we need to determine critical properties and behavioral characteristics. Perfecting and using new techniques, such as remote sensing and computerized data collection, can reduce costs. In addition, the number of field experiments and data analyses can be reduced by establishing a well-distributed network of sampling and study areas in each state, by standardizing experimental procedures, by presenting background information from study areas in a uniform format, and by extrapolating properties and behavioral characteristics of soils.

3. Integrated System to Handle Research Data on Soils

A wealth of information exists on soil properties and performance characteristics for thousands of soils, but it cannot be used effectively because it is scattered among numerous publications and reports. A computerized data bank should be developed that is capable of organizing and presenting all pertinent facts for soils identified by the National Cooperative Soil Survey.

4. Models for Reacting to Forces Outside U.S. Agriculture

The U.S. Government cannot accurately predict consequences of forces outside the agricultural sector and the nation on uses of resources in agricultural production, processing, and marketing. Population growth, foreign military actions, and manipulations of such world commodities as oil and grain are examples of external forces. Research must develop mathematical models that can help analyze these forces and guide policies on land and water use.

5. Integrated Land Information System

Land resource data that describe the physical, cultural, and ownership characteristics of land need to be assembled and maintained for all parts of the United States. Technology for data collection through a combination of satellite, aircraft, and ground-based acquisition systems is available. However, a system must be developed to monitor land use and changes in land qualities with the precision necessary for resource planning and management.

Research is needed to improve sensor capabilities, data processing, and integration of procedures to handle information from various sources. A system for handling cadastral (land ownership) data needs to be developed that will achieve compatibility between data generated by various units of government. Agreement must be reached on the frequency and means by which data are collected to maintain a land-related information system.

REFERENCES

1. Conservation Needs Inventory Committee. 1962. Basic statistics of the national inventory of soil and water conservation needs. Statistical Bull. No. 317. USDA, Washington, D.C.
2. Dideriksen, Raymond I., Allen R. Hidlebaugh, and Keith O. Schmude. 1977. Potential cropland study. Statistical Bull. No. 578. SCS, USDA, Washington, D.C.
3. Larson, W.E. 1981. Protecting the soil resource base. J. Soil Water Conserv. 36:13-16.
4. National Research Council. 1980. Need for a multipurpose cadastre. p. 1-4. In National Academy of Sciences, Washington, D.C.
5. Schmude, Keith O. 1977. A perspective on prime farmland. J. Soil Water Conserv. 32:240-242.
6. U.S. Bureau of the Census. 1970. Areas of the United States. E-20, No. 1. U.S. Department of Commerce, Washington, D.C.
7. U.S. Council on Environmental Quality and Department of State, 1980. The global 2000 report to the President. Tech. Rep. Vol. II. U.S. Government Printing Office, Washington, D.C.
8. U.S. Department of Agriculture. 1961. A universal equation for measuring wind erosion–An aid to conservation farming in the Great Plains. Agricultural Research Service Special Rep. 22-69, Washington, D.C.
9. U.S. Department of Agriculture. 1971. Basic Statistics–National inventory of soil and water conservation needs, 1967. Statistical Bull. No. 461, Washington, D.C.

10. U.S. Department of Agriculture. 1979. Soil and water resources conservation act: Appraisal 1980. p. 3-35 to 3-44. Review Draft Part I. Washington, D.C.
11. U.S. Department of Agriculture. 1980. An assessment of the forest and range land situation in the United States. p. 34–35. FS-345. In U.S. Forest Service, Washington, D.C.
12. Wischmeier, Walter H., and Dwight. O. Smith. 1978. Predicting rainfall erosion losses. USDA Agric. Handb. No. 537. U.S. Government Printing Office, Washington, D.C.

3

David M. Van Doren, Jr., Chairman

Harold L. Barrows, Orvis P. Engelstad,
G. R. Foster, Ervin J. J. Koos,
George W. Langdale, Robert F. Lucey,
Jerry V. Mannering, William P. Martin,
Z. B. Mayo, Paul G. Orth, Gary Osborne,
James A. Ryan, W. D. Seitz, Albert E. Smith, Jr.

HUMID REGIONS

Conservation Needs and Technology on Agricultural Land

In 1977 approximately 293 million acres of land were devoted to agriculture in the humid region of the USA. Millions of additional acres had been irreversibly removed from production during the last several decades because of excessive erosion and conversion to other uses. In order to meet increasing demand for agricultural products, increased ability to protect production resources is required. This paper describes current conservation resource conditions in the humid region and presents a prioritized list of interrelated research items whose development and adoption should provide most of the needed resource protection.

Humid regions are geographic areas where satisfactory crop yields can be produced in years with normal precipitation without resorting to irrigation or fallow periods specifically designed for conservation of water. In the United States the humid region is east of the 96 degrees west longitude (Fig. 1). Agricultural lands are natural landscape areas devoted to production of food, feed, oil, and cultivated fiber crops. The term natural landscape does not include drastically altered landscapes such as surface mines. Soil and water conservation technologies are those production practices and special structures used to control soil and water movement across or off land.

Application of technology which satisfies demand for agricultural products from fewer acres permits placing some land in reserve or at least permits placing land with high erosion potential in uses less subject to damaging erosion (pasture, forest). Improvements in equipment, plant

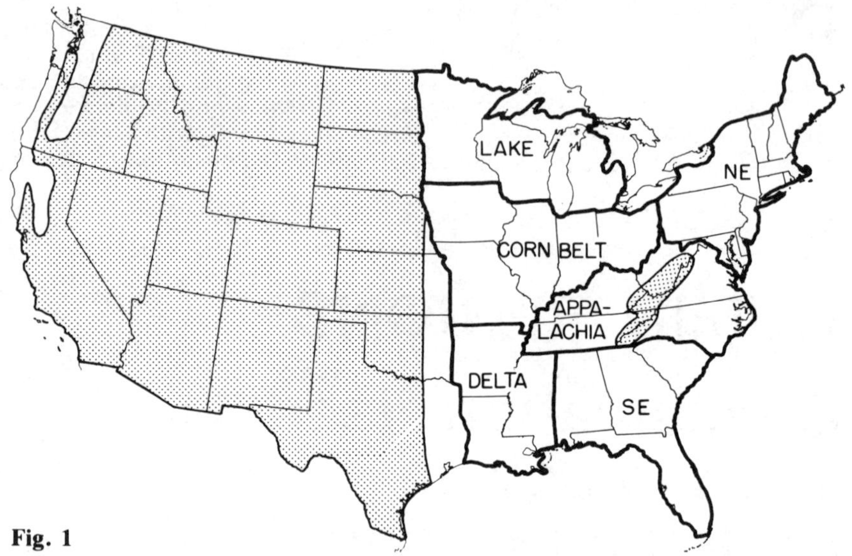

Fig. 1

materials, chemicals, or general management, even if not related to conservation technology, may provide this relief. While appropriate to achieving the goals of conservation technology, these improvements will not be considered in the process of developing prioritized research needs.

The basis for deciding future technology needs is an accurate inventory of current resources (land, water, climate, and technology) and current pressures for and constraints on use of resources. In addition, prediction of trends in inventory and resource use as well as projected impacts of new technology are important in deciding the fate of soil and water resources. The problems to be solved then become those instances where the resource inventories do not or are not expected to satisfy the demands of one or more of the pressures or constraints.

CURRENT SITUATION

Soil and Water Resource Base

The approximate land area in the humid region devoted to agriculture in 1977 is shown in Table 1. Intershifting among cropland, pasture, and forest has occurred during the past few decades and will continue. However, the total land available for cropland and pasture in the humid region is continually being irreversibly reduced by conversion to uses for transportation, urban development, and water (lakes, ponds). Erosion has also removed some land from agricultural production and has reduced productivity of other land to an as yet unmeasured extent. Erosion occurred at an estimated rate of 5 tons/acre or more per year on 27 percent of 1977 cropland and pasture (Table 2). Such an erosion rate is generally considered great enough to impair or reduce productivity. However, accurate data for the assessment of this impact of erosion on productivity are not available.

The soil resource base is not of uniform quality. Approximately 80 percent is rated as land use capability Classes I, II, or III (USDA, SCS, 1978). The latter two classes have some restrictions of wetness, erodibility, stoniness, or droughtiness. All three classes, generally with

Table 1. Distribution of 1977 cropland and pasture in the humid region and land potentially convertible to cropland and pasture (adapted from USDA, SCS, 1978).†

Land use	Subregions						
	Corn Belt	Lake	Appalachia	Delta	SE	NE	Total
	million acres						
			1977				
Cropland	89.8	44.1	20.7	21.2	17.5	16.9	210.2
Pasture	25.2	6.9	18.5	12.6	14.1	5.8	83.1
Total	115.0	51.0	39.2	33.8	31.6	22.7	293.3
			Future				
Cropland‡	15.5	9.3	15.2	10.6	17.1	5.7	73.4
Pasture§	18.3	23.6	20.9	22.3	35.4	25.1	145.6

† Cropland is agricultural land harvested primarily by humans. Pasture is agricultural land harvested primarily by livestock.

‡ Future land which has high and medium potential for conversion to cropland contains significant percentages of land currently used as pasture. Do not add total 1977 land to future land convertible to cropland.

§ Future land which has low potential for conversion to cropland but which may have a reasonable potential for conversion to pasture, if not already so used.

Table 2. Annual soil losses at 1977 rates from cropland and pasture in the humid regions (adapted from USDA, SCS, 1978).

Land use	Soil erosion class	Land area in each erosion class							Total estimated annual soil loss†
		Corn Belt	Lake	Appa-lachia	Delta	SE	NE	Total	
	tons/acre	million acres							million tons
Cropland	<2	25.6	29.0	7.1	2.6	5.2	9.2	78.7	78.7
	2–5	30.3	9.4	5.4	7.5	5.4	3.5	61.5	215.2
	5–10	17.3	3.5	3.8	7.4	4.0	2.1	38.1	285.8
	>10	16.6	2.2	4.4	3.7	2.9	2.1	31.9	478.5
	Subtotal	89.8	44.1	20.7	21.2	17.5	16.9	210.2	1,058.2
Pasture	<2	16.7	6.1	12.0	9.6	13.2	5.0	62.6	62.6
	2–5	3.6	0.5	2.9	1.5	0.6	0.4	9.5	33.2
	5–10	2.6	0.2	1.7	0.7	0.2	0.2	5.6	42.0
	>10	2.3	0.1	1.9	0.8	0.1	0.2	5.4	81.0
	Subtotal	25.2	6.9	18.5	12.6	14.1	5.8	83.1	218.8
Total		115.0	51.0	39.2	33.8	31.6	22.7	293.3	1,277.0

† Estimated soil loss at 1 T/A per year for class <2; 3.5 T/A per year for class 2–5; 7.5 T/A per year for class 5–10; 15 T/A per year for class >10.

lower yields for higher class number, are capable of producing acceptable crop yields under continuous cultivation using generally accepted 1977 technology. The remaining 20 percent of the land resource base has such severe limitations, primarily erodibility, that only occasional cultivation is currently recommended.

Water supply for crop production in the humid region has historically been derived primarily from precipitation and water stored in the soil without recourse to irrigation. On the average, annual precipitation exceeds annual potential evapotranspiration, although along the western boundary of the humid region the situation is often reversed. Averages can be deceiving, however. In most growing seasons there are periods of 1 to 6 weeks when plant growth is reduced by lack of water. Evapotranspiration during this time of year is expected to equal or exceed precipitation for average years. Also, seasonal evapotranspiration, if water is available, does not vary much, whereas rainfall is extremely variable.

Much of the rain that does occur during the late spring and summer occurs as short, high intensity events, resulting in a high potential for runoff. Surface runoff may be beneficial for recreation and municipal and industrial supplies, but it is currently of little value to crop production. As an indication of variability of water supply for agriculture, annual runoff ranges from 10 to 300 percent of average annual precipitation over the humid region (USDA, 1980b).

The need for increased stability of crop production, long recognized in the semi-arid region, resulted in a 5.6 million acre increase in irrigated land from 1974 to 1977 bringing total irrigated acreage in the humid region to 23 million acres. Groundwater reserves are substantial, and as of 1975 were being reasonably well maintained in the humid region except for depletion in parts of Georgia, Illinois, Iowa, Louisiana, and Michigan (USDA 1980b). Greatly increased demand for irrigation water to the extent not satisfied by annual recharge of surface and groundwater supplies could eventually cause the kinds of problems encountered today in the semi-arid and arid region.

Demand for U.S. food and fiber products has increased markedly during the past decade. While domestic consumption of wheat and coarse grains has increased less than 10 percent each, exports have doubled and tripled, respectively. Production has been able to keep pace (Table 3) so that inflation-adjusted prices for these commodities has remained relatively constant. If domestic demand for biomass for conversion to liquid fuels becomes appreciable, domestic consumption of coarse grains (especially corn) could rise at rates considerably greater than during the past decade.

Agricultural production capacity in the United States cannot be considered at a point of immediate crisis. While significant annual fluctuations occur on a regional and less often on a national basis for a given commodity, the capacity is presently sufficient for current and near future demand. Some adjustment by the nation's farmers may need to occur in response to shifting demand among commodities. The land and water base should be capable of supporting the needed shifts. For example, potential future cropland (land suitable for cropland but currently used for other purposes) equals one-third of land presently used for cropland in the humid region (Table 1). Most land used for corn could be used for soybeans and vice versa. Potential future pasture land (land with low potential for cropland but currently used for other purposes) probably exceeds that now used for pasture. Productivity might be changed up or down somewhat, depending on the particular use changes, but a large capacity for shifting does exist.

Prospects for 10 to 20 years hence are not so sanguine. World demand for U.S. agricultural production will continue to increase. Water shortages for irrigation in the semi-arid areas of the U.S. may put more of a demand on humid area resources to make up for the lost production. Large scale, intermittent irrigation can be used to supplement precipitation as long as water supplies are efficiently managed. Land will continue to be irreversibly lost from agricultural production to roads, lakes, and urban developments. While additional lands can be shifted to agricultural production, the quantity is finite. If erosion continues at the present rate, considerable areas of the humid region will suffer productivity loss. The social, environmental, and economic problems caused by soil and chemicals removed from the land by runoff and erosion plus chemical movement by leaching to the ground water may exceed the economic problem of reduced land productivity.

Significant research results and technological advances generally accrue from sustained efforts. Researchers and research facilities cannot be assembled overnight. Technology is not disseminated or adopted in a few years. Consequently, in order to develop and adopt the technology needed 20 or more years hence, the efforts must start now. The lack of a current crisis gives the research community the opportunity to be thorough and much more efficient than if it were operating under the pressure of an imminent shortage of food.

Table 3. Average land area, production, and export of wheat, coarse grains, soybeans, cotton, and hay for 1977/78 through 1979/80 crop years (adapted from USDA, 1980a).

	Geographic area	All wheat	Coarse† grains	Soybeans	Cotton	All hay
Land area in million acres	Humid USA	11.5 (0.02)‡	63.8 (0.07)‡	61.2 (0.50)‡	3.6 (0.04)‡	31.4
	Total USA	62.8 (0.11)	106.7 (0.12)	64.5 (0.53)	13.0 (0.16)	61.1
	World	565.2	870.5	122.5	80.8	--
Production in million tons	Humid USA	13.0 (0.03)	171.2 (0.21)	55.2 (0.60)	0.9 (0.06)	73.9
	Total USA	59.6 (0.13)	240.7 (0.30)	58.9 (0.64)	3.2 (0.21)	139.8
	World	458.4	801.0	92.2	15.2	--
Export in million tons	Total USA	37.3 (0.44)	66.2 (0.66)	23.0 (0.83)	7.6 (0.33)	--
	World	83.9	100.0	27.7	4.9	--

† Coarse grains include corn, sorghum, millet, oats, barley, and rye.
‡ Values in parenthesis are fractions of the world values in the same category.

Agricultural Production Technology Base

The specific technology base differs among groups of crops. Annual crops differ from perennial crops, row crops from solid seeded crops, grains from vegetables. The discussion to follow will be somewhat general, without describing each type of technology for every crop grown in the humid region. Specific examples will be used where appropriate to illustrate a point.

1. Land Base Improvement and Protection

Erosion control technologies for protection of the land base have been developing continuously since the awareness of need brought on by the Dust Bowl in the mid–1930's. Today some form of control measures are used by many farmers in the humid region. These include field production practices of crop-residue management, use of cover crops, rotations, contour and strip cropping, ridges, and terraces. Also used are special non-production technologies such as grassed waterways and permanent structures along waterways.

Unfortunately, not enough of these control technologies are in place. If it is assumed that erosion at rates greater than 5 tons/acre per year are damaging to future productivity of any soil, at least one-third of the cropland in the humid region was suffering damaging erosion in 1977 (Table 2). If, as some data indicate, erosion rates greater than 2 tons/acre per year are damaging to any soil, almost two-thirds of the cropland was

Table 4. Estimated annual soil losses at 1977 rates from cropland eroding at rates greater than 5 tons/acre per year in the humid region (adapted from USDA, SCS, 1978).

	Estimated soil losses						
Crop	Corn Belt	Lake	Appa-lachia	Delta	SE	NE	Total
	million tons						
Corn	267.2	57.2	41.9	6.9	36.8	33.5	443.5
Soybeans	147.3	14.4	44.7	84.2	30.9	5.7	327.2
Other	95.9	49.6	33.4	48.9	29.9	29.6	287.3
Total	510.4	121.2	120.0	140.0	97.6	68.8	1,058.0

undergoing excessive erosion. Estimated total annual soil loss from lands eroding at rates exceeding 5 tons/acre per year are shown in Table 4. Land in corn and soybeans is the source of 75 percent of this total erosion. Sediment resulting from erosion on cropland may degrade off-site water quality and reduce the capacity of water conveyence structures, reservoirs, and harbors.

Current promising technology is based on maintenance of a cover of plant material, alive or dead, on the soil surface at all times along with special conservation tillage techniques. Unfortunately, the costs of such technology in its present state of development are perceived to be too great for mass adoption in today's economy. These "costs" may include reduced crop yields and inefficient use of chemicals in certain incompletely understood circumstances, different equipment needs, greater reliance on pesticides, and a general change in production practices. Advantages to some of this technology in addition to controlling erosion within a 2 to 5 ton limit are potential labor savings of up to 40 percent for corn and soybean production, potential energy savings, and improved water use efficiency during the growing season for summer annuals. Technology does not yet exist to permit cropland use as in 1977 without exceeding a 2 ton/acre per year limit for much row crop land. If such a limit were required today for maintenance of soil productivity or to meet other societal goals, row crop production would be severely limited and would not meet current demands.

2. *Equipment*

The trend in U.S. agriculture has been toward larger farms, equipment, and greater labor efficiency. Production per hour and acres farmed per hour have both increased substantially during the past four decades. Recommendations of technologies which entail increases in labor input per acre will be poorly received.

The increased size of production and harvesting equipment has unquantified negative effects. The increased load from heavy equipment tends to increase soil compaction, especially since these machines can be operated on wetter soils that are highly susceptible to compaction. Soil compaction reduces pore space, decreases rate of drainage, and increases soil density which often results in decreased root growth. This is recognized as a significant limiting factor in current crop production in many parts of the humid region.

Much of the current equipment used for planting, cultivation, and chemical application was designed to operate in residue-free fields. The modification of this equipment to operate in residue has been reasonably satisfactory, but improvements would speed adoption of the newly developing conservation-tillage technologies.

3. *Plant Materials*

Plant breeders have developed improved varieties of agricultural crops that produce greater yields, have greater resistance to lodging, are more resistant to insects and diseases, produce well with greater numbers of plants per acre, produce material of better quality, and mature more uniformly to improve mechanical harvesting. Improved varieties are estimated to have been responsible for about half of the yield increases achieved in grain production in the United States during the past 20 years.

These improvements have been achieved within the general environment of residue-free fields. That is, most or all of any remains of the previous crop have been removed from the soil surface by removal from the field (as in making silage from annual row crops) or by burying in the soil with tillage implements. One of the most promising technologies for reducing soil erosion is to have the soil surface covered with crop residues. This technology changes the soil environment for the plant (soil temperature, density, water content) and also for insects, diseases, and weeds. Perhaps if varieties had been selected during the past 20 years within the general environment of residue-covered soil, the resulting array of recommended varieties would have been more compatible with this type of conservation technology.

4. *Chemicals*

Use of fertilizer, lime, and pesticides has increased dramatically during the past four decades. Greater yields per acre associated with other technological improvements have required greater inputs of fertilizers in order to replace nutrients removed during harvest or lost by erosion, runoff, leaching, or other mechanisms. Decline in the number of animals used for farm power and the concentration of the livestock industry have reduced the supply of manure formerly used for fertilizer. Reduced planting of forage legumes which "use" nitrogen from the air has decreased this source of nitrogen for succeeding crops.

Despite recent large increases in cost of fertilizer, use has continued to increase. The tendency to apply excess fertilizer may reverse if prices continue to rise as in the past 5 years, and controls governing water quality become more stringent. Evidence shows reserves of phosphate and potash built up by repeated fertilization are available to crops and can support high crop yields for several years without further addition. It is also possible that if crop yields are to reach 50 to 100 percent higher levels than current averages, even greater quantities of nitrogen, phosphate, potash, and other nutrients will be required.

Present formulations of fertilizers were developed when prices were low and there was only limited concern for losses of chemicals from land. Future fertilizer technology must consider efficiency of use by crops. Means of achieving this are altering of chemical and physical form, improving application timing and placement, and using other means for preserving availability of applied nutrients for the intended crop. In addition, this technology must complement technologies for control of erosion and nonpoint source pollution.

Pesticide use has increased for many reasons. More complete weed control using less labor can often be achieved at lower total cost per acre with herbicides than by using strictly mechanical means (tillage, burning). Under residue-free conditions, a complete lack of mechanical cultivation following seedbed preparation can result in greater water runoff and subsequent lower yields on some soils. However, a combination of chemical and mechanical weed control can be profitably accommodated. Some forms of conservation-tillage practices depend solely on chemicals for weed control. Some weed species are controlled better with chemicals than by mechanical means under many circumstances. Herbicides are continually being developed that have better broad spectrum control while not damaging the crop or succeeding crops, and which need not be incorporated into the soil. As this trend continues, flexibility of other production technologies will be improved.

Since awareness of potential hazards associated with pesticides has increased, many effective materials have been removed from approved use and the cost of introducing new materials has skyrocketed. Consequently, the arsenal of materials available for use for pest control is somewhat more limited than might be expected, and the development of new materials adapted to the emerging conservation-tillage technologies will be slow.

Integrated pest management (IPM) holds promise for more effective use of pesticides. Where appropriate, IPM attempts to control pests with combinations of pesticides, biological controls, pheromones, traps, cultural practices, and any other practice that will hold populations of the pest to less than economically damaging levels. This new phrasing of an old concept is considerably hampered by lack of knowledge regarding interaction of pests with their environments, particularly soil and residue-inhabiting species. Should IPM develop and reduce demand for pesticides, newly developed or improved conservation tillage technologies will have one less obstacle to overcome for acceptance.

5. *Water Management*

Annual precipitation is adequate for crop growth in the humid region. Unfortunately, the distribution of precipitation throughout the growing season is not uniform. There are times during the year when the soil is too wet, and often during the growing season there are periods of moisture deficit that adversely affect crop growth and yield.

Many farmers have attempted to overcome these problems by installing drainage systems to remove excess water, and irrigation systems to provide supplemental water. Indeed, there are about 100 million acres artificially drained in the humid region. Almost without exception, the output from the drainage system is directed to the nearest stream. No provision is made for storing the water for use during the dry periods.

Excess surface and soil wetness often hinders timely operations such as tillage, planting, chemical application, and harvesting. Excessive wetness drowns out crops during germination or early growth. Proper surface and subsurface drainage is essential to reduce crop production losses in years when precipitation is excessive.

Conservation tillage systems where plant residues are maintained on the soil surface can lead to reduced runoff and evaporation of precipitation. The residue, if in sufficient quantities, can alleviate the effect of raindrop impact on crusting or sealing of the soil surface. This tends to increase or maintain infiltration, reduce runoff of water and chemicals, and assist in the protection of the crop from variabilities of climate and, thus, water stress. No-till, a form of conservation tillage, with high mulch levels can lead to improvement in natural internal drainage by the production and maintenance of continuous pore spaces in some soils.

6. *Soil and Crop Management*

Agricultural production requires not only the basic land preparation and protection technologies, equipment, plant materials, and chemicals, but also the skills to utilize these factors for optimum production. Proper timing of field operations is essential for economic production. Improper timing can largely negate potential benefits of any technology. This extremely important aspect of agricultural production is difficult to master especially when dealing with the uncertainty of weather, which affects the preparation of seedbeds, application of chemicals, and harvest of crops. The more successful new technologies developed for soil and water conservation will be those providing the greatest flexibility in the timing of production operations. In other words, the successful conservation technologies will provide maximum insulation from variability of weather.

Technology which permits direct substitution of a low erosion potential use for a high erosion potential use will also improve total conservation of the land base (such as substituting high quality forage for feed grain for cattle). The challenge is to be able to insure sufficient management so that the low erosion potential use does result in acceptable soil losses.

External Pressures for and Constraints on use of Agricultural Production Technologies

Farming, like any other private industry, requires profit to continue in business. After adjusting for inflation, net farm income has increased about 15 percent in the past two decades while prices for farm commodities (primarily grains) have remained constant. This income improvement would not have been possible without economies of scale (larger acreages operated with less labor) and adoption of improved production technologies, which reduced unit costs of commodity production. Proposed soil and water conservation technologies which increase unit costs will not be adopted without external incentives. In fact, many formerly used conservation technologies such as contour tillage, terraces, strip cropping, and crop rotations are being abandoned without regard to their long-term soil-conserving benefits. The associated increase in soil erosion is of little concern to many farmers because they are not convinced of the cost to them in reduced yields resulting from erosion. Only soil deposition or gullies which "visibly" hinder normal operations are considered important in many cases.

In the past four decades improved agricultural production technology generally relied on use of cheap, abundant energy and larger amounts of various chemicals. Prices of energy and some chemicals have risen faster than the general rate of inflation during the past 7 years. For the foreseeable future, energy supplies, while not necessarily abundant, will be sufficient, but the price may be expected to rise far more rapidly than the general inflation rate. Supplies of phosphate rock for phosphatic fertilizer should also be ample. However, the use of increasingly lower grade rock will exacerbate rapidly increasing price trends. Tightness of world sulfur supplies also has led and will continue to lead to increasing costs and prices of phosphatic fertilizers. Potash supplies are also ample. However, worldwide shortages of effective mining capacity for potash fertilizers will drive prices up until new effective mining capacity becomes operational. Dwindling supplies and increasing prices will result in the preference for soil conservation technologies efficient in the utilization of these resources.

Social and political pressure may ultimately provide the greatest incentive for improved soil and water conservation technology. Water quality standards for the nation's surface and ground-water supplies are becoming more stringent. Less sediment beyond farm borders is being tolerated for various economic, aesthetic, political, environmental, or social reasons. Thus, adequate control of nonpoint source pollution due to erosion from cropland may require lower soil-loss tolerances than are necessary to maintain long-term soil productivity.

Less easily satisfied demands by society may be in the area of reduction in chemical transport off the farm. The trend toward increased use of chemicals will make such goals more difficult to achieve without increasing unit production costs. Also, technologies which reduce soil loss do not necessarily reduce losses of chemicals, especially the more soluble ones, to the same extent that soil losses are reduced. New technologies may be needed instead of merely improved current technologies.

For the past 40 years, society has recognized that it has a real stake in continued productivity of the land. Cost-sharing programs and funding for advisory services have helped farmers defray costs for adopting specific soil conservation technologies. To the extent that this willingness to pay endures, the potential for adoption of such technologies will be enhanced over the potential based on purely personal economic considerations.

STATE OF THE PROBLEMS

Priorities for research to improve soil and water conservation technologies fall into five important groups. The groups and problems within groups will be listed in order of priority.

A. Improve the ability to grow crops on soil surfaces nearly covered with plant materials (alive or dead). on the current inventory of productive agricultural land using annually applied technology.

The land area now used for agricultural production needs immediate protection from soil erosion in order to preserve long-term productivity and to satisfy demands for improved water quality (e.g., reduced non-point source sediment and chemical movement into surface waters). Recent evidence indicates that "T" values used to estimate the maximum rate of annual erosion that will not cause long-term productivity decline are perhaps too high (personal communication with Dr. G. F. Hall, Ohio State University, Columbus, OH 43210). Therefore, the nation is, in effect, mining its productive cropland base in the humid region. Current generations can survive nicely; but future generations cannot.

Large soil losses (greater than 5 tons/acre per year) from cropland, such as corn and soybean acreage in the Cornbelt will not long be tolerated by society. If allowed to continue, the farmer will eventually be forced to pay directly for such excessive erosion by reduced returns, or he will pay indirectly by being forced to adopt conservation measures that may further reduce net income. If those conservation measures result in lower yields per acre or in fewer acres in production, consumers will pay through higher prices caused by lower supplies of commodities.

Therefore, it is imperative that effective, economical soil conservation technologies be available for adoption by farmers. Society is faced with a problem because much of the currently available technology is considered too expensive to use. Contour farming requires time and care in keeping operations close to the contour and in handling irregularly shaped end-rows. Strip cropping and rotations, particularly on cash crop farms, are not generally accepted because they either require the use of crops (oats and barley) that currently return to farmers a lower dollar profit compared with corn and soybeans or they require crops (forages) not needed or desired by the individual farmer. Seeding of cover crops takes time and requires expense for seed and fuel. Grassed waterways require care to maintain and to avoid damage while tilling the land. One of the newer conservation-tillage technologies (no-till) requires increased reliance on expensive chemicals or relies too heavily on specific climatic conditions to make chemical control effective. Although undeniably the most effective erosion-control technology when used properly, no-till may result in lower yield. Even if some of the above technologies (contouring, strip cropping, rotations) were largely adopted, they may inadequately control erosion if high production from row crops is to be maintained (Table 5).

In addition to reduced costs for any improvement, achieving this objective will require the following:

1. Determine the best practical combinations of soil physical conditions combined with residue cover for potential crop production as well as soil and water conservation. This includes evaluation of the effects of conservation tillage methods which leave appreciable amounts of residue on the surface on those soil properties that govern water runoff, infiltration, and internal drainage (soil aggregation, pore distribution, and continuity).

2. Determine the best combination of techniques for establishing and maintaining satisfactory vegetative cover on the soil surface. This includes equipment for distributing residues as desired on the surface

(with or without tillage) and techniques for establishing and maintaining cover crops. Vegetative cover is currently the best single control of soil erosion (Table 5). This objective is particularly relevant to removal of large quantities of plants and plant residues for feed or fuel. The amount of vegetative material that can be removed is limited in order to keep adequate residue for prevention of soil degradation. Otherwise, a supplemental cover crop is required to provide the necessary soil cover.

3. Develop technology for insuring adequate stands of desired species. This includes improved varieties which are more vigorous in cool, moist conditions under vegetative cover and equipment to plant at higher speeds and within narrow depth limits in the presence of vegetative cover, especially in moist or high clay content soils. Reduced or erratic stands, especially if they occur at decreased planting speeds, are major deterrents to adoption of any technology.

4. Develop improved chemical pest control technologies in the presence of cool, moist, residue-covered, and high surface soil organic matter content conditions. In addition to the need for a broader spectrum of chemicals with greater reliability, these technologies include equipment for applying materials accurately and efficiently (under residue, directed spray), reduced need for incorporation, reduced phytotoxicity with seed and seedlings, greater residual effects (especially for insecticides), and greater safety for succeeding crops (herbicides). Unreliability of weed control has been a major deterrent to adoption of current conservation tillage technologies. Insects and diseases have been a lesser deterrent.

5. Develop technologies for improved fertilizer efficiency under conditions of conservation-tillage systems, multiple cropping, and supplemental

Table 5. Relative effectiveness of various erosion control technologies (adapted from Bone, 1979).

Technology	Relative erosion control†
Continuous cropping (corn, soybeans); residues removed; plowed	1.00
Continuous cropping (corn, soybeans); residues left; plowed	0.75
Contour tillage (multiply by other technology)	0.5 to 0.9‡
Terracing (multiply by other technology)	0.5 to 0.67§
Crop rotation (corn, soybeans, small grain); residues left; plowed	0.4
Crop rotation (corn, soybeans, small grain); residues left (half on surface)	0.25
Continuous cropping (corn); residues left (all on surface); not plowed	0.1
Continuous meadow; residues left; not plowed	0.01

† Smaller numbers represent proportionally lower annual erosion rates.
‡ Values increase as slope steepness increases.
§ Values increase as slope steepness and/or slope length decrease.

irrigation. In each case, fertilizer is usually surface applied, resulting in a buildup of the more immobile nutrients (phosphorus and potassium) at or near the soil surface. These nutrients are available to crop roots only to the extent that moisture also is present in that layer. Nitrogen, especially in the form of urea or urea-based materials, is subject to volatilization losses with surface application. In addition, the nitrification of surface-applied ammonium-N produces an acid surface layer, which, in turn, reduces the effectiveness of the triazine herbicides. Technology should be developed to allow application of fertilizer materials below the residue-covered soil surface. In some cases, new or chemically altered nitrogen sources should be formulated for these systems.

Also, studies of fertilizer requirements under these new tillage systems are needed. The cooler temperature and higher moisture levels generally associated with residue-covered soils can be expected to alter plant-nutrient behavior. Fertilizer recommendations may need to be specifically developed for these conditions.

6. Develop cropping systems which involve inter-row legumes and grasses. Such systems offer alternative control techniques for the conservation of soil and water. In addition, a reduction in energy use may result from reduced nitrogen fertilizer inputs when legumes are included.

7. Improve efficiency of non-chemical pest-control techniques. The influence of cropping systems on the biology of pests and beneficial species is not understood. These relationships are especially important to the development of viable pest-control programs. Reduction in use of pesticides would reduce potential for movement into surface or subsurface waters. However, such movement should be a minor deterrent to adoption of conservation technology if soil erosion control is coupled with reasonable runoff reduction.

B. Improve the capability to inventory and predict the effect of climate, soil, topography, land use and management on soil productivity and soil degradation from erosion, compaction crusting, and loss of organic matter and nutrients.

Gross production from the nation's farmland has increased more or less steadily for decades. Average statewide records for per-acre production of most grains are broken regularly. This production increase is masking a slower decrease in production potentials. At the present time there is no universally accepted method for determining the actual production potential, much less the rate of decline. Lacking such capability, it is impossible to determine the extent of the damage being done by current rates of erosion or by any other process. Until such capability is achieved, national priorities and strategies related to soil conservation based on preserving the productivity of the soil are tenuous at best. The capability to predict sheet and rill erosion is better developed. The widely used Universal Soil Loss Equation is based on empirical relationships developed from thousands of plot years of data. While this provides reasonable estimates of average annual erosion from given circumstances of location, soil, and management, it does not describe the rate or quantity of movement off the field or farm. Delivery of sediment to waterways, and movement of soil in complex watersheds are weak links in the ability to develop national priorities and strategies related to protection of quality of surface waters.

Achieving the objectives will require:

1. Development of physiologically and physically based models for accurately simulating growth of the major crops grown in the humid region as a function of climate, soil, and management inputs. Such models should be sensitive to changes in soil profile characteristics associated with soil loss by erosion and bringing subsoil material to the surface by tillage. Then, assuming that a suitable data base were available, computations could describe productivity changes for any shift in land use or management or for any degree of soil degradation.

2. Development of a comprehensive erosion-transport-deposition model to account for all sources and modes of transport of sediment in the landscape. This includes determination of basic chemical and physical processes taking place when soil particles are detached, transported, and deposited, as well as management alternatives which would alter the processes. The model should cover a wider range of slope lengths, steepness, and soil textures than are now covered by the Universal Soil Loss Equation. Once such a model were operable, soil loss from fields could be estimated for current or projected erosion-control technologies to use with soil productivity computations to better assess changes therein. Also, assessment of needed erosion-control strategies for protecting water quality could be more accurately determined on a local or national level.

3. Development of a transport model to describe movement of soluble chemicals in surface runoff, tile flow, soil water, and groundwater, plus movement of soil–absorbed chemicals on sediment leaving farm fields and agricultural watersheds. Such a model is required to assess the impact of agricultural management on off-site water quality and to guide the selection of on-farm, site-specific practices to prevent excessive water-quality degradation.

4. Improvement and extension of the data base for developing, validating, and implementing erosion and sediment yield models. Data are necessary for developing basic relationships for the controlling processes, for validating the models to insure that they give acceptable results, and

for providing input values required to use the models in their intended applications. Collection of these data will require a variety of experiments ranging from basic laboratory research to studies involving field plots, small watersheds, and large complex areas. Collection of data is very expensive, and great care must be exercised to insure accurate and appropriate data that satisfies multiple needs.

C. Improve Water Management for Efficient Crop Production

Permanent improvements for water management and soil conservation (surface and subsurface drainage systems, irrigation, terraces, and waterways) while lasting from 10 to 50 years, require a large capital investment generally payable during the first third of the expected lifetime. The expense has deterred many from using these technologies, especially if these technologies cannot be shown to have a favorable benefit-cost ratio. Improved drainage and supplemental irrigation generally increase yields and allow for some cropping practices that would otherwise be impossible. Irrigation would not be economical with today's economy for the majority of land in the humid region. However, for those soils and crops where irrigation is beneficial, information on water supply, plant water requirements, costs, and management is needed. Improved water management allows for more timely and efficient field operations.

Terraces, waterways, and grade-control structures support other conservation practices such as conservation tillage to prevent excessive erosion in waterways and gully formation which inconveniences farm operations, removes land from production, and reduces land value. Impoundment type terraces are especially effective for control of sediment loss from fields.

This objective could be met by:

1. Determining the drainage requirement for important crops using a range of conservation technologies which maintain vegetative cover (alive or dead) on the soil surface all or part of the time or which rely on soil roughness to provide protection from erosion. If erosion-control requirements become sufficiently restrictive that erosion control is needed on naturally poorly drained land, drainage systems may need to be designed specifically to produce a more rapid drainage than is required for residue-free technologies in order to avoid yield losses. Research may also be required to determine how such levels of drainage might be accomplished at a reasonable cost, especially on the hard to drain soils having very shallow fragipans or very high clay contents.

2. Determining how terrace and waterway construction can be accomplished most economically and with as little loss as possible of

productive land under humid conditions in relatively complex topography. If this objective could be accomplished, it would provide an alternative or supplement to technology developed in part A.

3. Developing technologies for efficient irrigation in the humid region. These technologies include systems for collecting and storing water during periods of excess precipitation, quantifying daily and seasonal crop water requirements, real time irrigation-scheduling technology to predict impending deficits, improved drainage systems that may be reversible where water supplies are available for irrigation, irrigation systems that are energy efficient and capable of alleviating short periods of drought but applying limited amounts of water to reduce adverse effects of excess water if heavy rains occur after irrigating, modified soil-management practices to enhance rooting depth, and plant breeding to develop crops tolerant to wide fluctuation in available soil water. The improved production and time-stability of production would benefit the indivdual farmer and would provide a more stable price structure for the rest of society.

D. Improve the ability to reclaim land damaged by erosion or other misuse (strip mining excepted) use in crop production.

Several millions of acres of formerly productive cropland have been lost to production because of complete removal of topsoil and resulting reduction in productivity. Gullies, which make farming inconvenient or impossible, have also taken cropland out of production. If, in the next century, there is sufficient demand for cropland which cannot be converted from other uses, such reclamation may be desirable, even at costs considered prohibitive today. An awareness of the extent to which such reclamation is or is not possible (including costs) may provide additional incentives for protection of the current inventory.

E. Improve technologies for converting land use as needed among forest, pasture, and cropland.

Shifting land use from cropland to pasture or from cropland and pasture to forest is generally no problem. Such shifts have routinely been made, especially if cropland or pasture has been damaged sufficiently by erosion to no longer be profitable for those uses. However, the reverse process, bringing forest land into use as cropland or pasture in the context of stringent erosion control and low cost during transformation, is not so easily accomplished. In most cases forest land has been converted directly to cropland. Perhaps it would be profitable to go from forest to pasture, which could then release some pasture land for conversion to cropland. Environmental impact resulting from these land-use conversions must be determined, and economical technologies for these shifts must be developed while maintaining a complete cover of vegetative material during and after the transformation. Perhaps the most difficult aspect is to make the shift economically attractive. Once such technologies were developed, the nation would have an improved capability for safely achieving any balance of land use desired at some time in the future.

RESEARCH PRIORITIES

Soil and water conservation technologies are those production practices and special structures used to control soil or water movement across or off the land. The humid region is that geographic area of the United States in which satisfactory crop yields can be produced in years with normal precipitation without resorting to irrigation or fallow periods specifically designed for conservation of water. In 1977, approximately 293 million acres of land were devoted to agriculture in this region; however, millions of other acres were irreversibly removed from production during the previous decade by conversion to other uses. Land has also been lost to erosion, but increased production on remaining land due to improved equipment, chemicals, plant varieties, and farm management has mitigated the effect.

Because of favorable climate and rainfall, the humid region is the key to sustained increased agricultural production in this country. Present technology for increasing production and stringently controlling erosion is often not accepted because of economic constraints. To develop and adopt the technology needed 20 or more years hence, we must start now. Research needed to develop the technologies to efficiently expand food and fiber production, while maintaining soil productivity and water quality in the humid region is outlined below:

1. Conservation-Tillage Technologies to Sustained Soil Productivity

Conventional tillage systems have caused excessive erosion on many productive soils, impaired water quality, and reduced the long-term viability of agriculture. Development and adoption of reliable, cost-effective, and energy-efficient conservation tillage technologies can reverse these trends and extend the nation's ability to satisfy demands for food and fiber.

2. Soil Productivity and Water Quality—Methods to Analyze Impacts of Agricultural-Management Systems

New knowledge and improved models of soil erosion, crop growth, and movement of water, sediment, and chemicals over and through the soil are needed to evaluate alternative management strategies for their potential impact on both short- and long-term soil productivity and off-site water quality.

3. Water Management Practices for Improved Soil and Water Conservation

Water management practices such as tile drainage, terraces, and irrigation are effective support practices, but their expense has limited their adoption. More cost-effective, land-conserving water management will increase crop production in the humid region.

4. Land Reclamation Technologies for Erosion-Depleted Soils

Erosion has reduced the soil productivity of millions of acres in the humid region. Accelerated food demands will require economical reclamation of these severely eroded lands.

5. Soil- and Water-Conserving Technology for Land-use Conversion

New technology is needed for minimizing soil erosion and maximizing water conservation during land-use changes such as from forest to pasture and cropland. Such technology will facilitate replacing acres shifted to non-farm uses; increase flexibility in choice of land use; and provide for increased demands for food and fiber.

REFERENCES

1. Bone, S.W., R. Christman, L.M. Feusner, R. Goettemoeller, and B.H. Nolte. 1979. Ohio Erosion Control and Sediment Pollution Statement Guide. Ohio State University. Coop. Ext. Service Bull 594.
2. Department of Agriculture. 1977. Changes in farm production and efficiency. USDA, ESCS, Statistical Bull 612.
3. Department of Agriculture, Soil and Conservation Service. 1978. 1977 National Resource Inventories.

4. Department of Agriculture. 1980a. Foreign Agriculture Circular. USDA, FAS, ESCS WCP-4-80.
5. Department of Agriculture. 1980b. Soil and Water Resources Conservation Act Appraisal 1980, Part I. USDA Review Draft.

4

Earl Burnett and R. John Hanks, Co-Chairmen

Raymond R. Allmaras, Jack J. Bond,
Donald G. Hanway, R. E. Highfill,
Marvin E. Jensen, Edward T. Kanemasu,
L. G. King, John Letey, Donald R. Nielsen,
Dick Piper, Francis H. Siddoway,
Melvin D. Skold

SEMI-ARID AND ARID REGIONS

Conservation Needs and Technology on Agricultural Land

The 11 contiguous Western States are classified as arid or semiarid (C.W. Thornthwaite, 1941, Atlas of Climatic Types in the United States, 1900-1939), except for parts of California, Oregon, and Washington. While the eastern parts of the Great Plains States are classified as either moist or dry subhumid, these areas experience many of the same adverse climatic extremes as the drier areas west of the 96th meridian. The arid and semiarid regions (17 Western States) are usually short of water for crop production without irrigation. Water deficits limit plant growth and crop yield to some extent almost every year.

Water deficits are not uniform or predictable from year-to-year. Crop yields are highly uncertain and substantial between-year variations in farm income are inherent to these arid and semiarid regions. A number of strategies have been devised and implemented which abate some of this uncertainty. Crop varieties have been developed which are more drought resistant; tillage machines and cultivation practices which conserve soil water are used. The crop-fallow practice reduces yield variability by storing some soil water for use in the subsequent year. Much of the nation's grain is produced in the vast semiarid dryland region of the country and yields on a per unit of water basis are comparable to yields in more humid regions. Yet, these strategies to meet uncertainty have often tended to subject the soils to additional erosion hazards or lead to other problems of resource conservation.

Extensive irrigation systems, developed to provide water for food and fiber production, have greatly increased agricultural productivity in the arid and semiarid regions. In 1974 irrigated land represented 18 percent of the total U.S. harvested cropland but produced 27 percent of the total value of agricultural products. Significantly most of the fresh produce grown in the USA comes from the irrigated southwest section of the country.

Water for irrigation has been derived from diverting surface waters and pumping groundwater. Some groundwater sources are naturally recharged at various rates, whereas others are essentially nonrechargeable. The Ogallala aquifer extending from Nebraska to Texas represents an example of the latter case.

The permanence of irrigated agriculture requires management to avoid excessive soil salinization. Some leaching is required to keep salinity levels in the root zone at acceptable levels, and artificial drainage systems are sometimes required to facilitate salt removal. Ultimately appropriate disposal of the excess salt is required. Leaching which is required to control soil salinity can also cause plant nutrients, particularly nitrate and various mobile pesticides to be removed from the soil profile. The leached nutrients and pesticides may be carried to groundwater sources or through drainage or seepage to reach surface waters. Thus excessive leaching is undesirable.

Irrigated and non-irrigated agricultural land in semiarid and arid regions contribute greatly to the U.S. food and fiber production, but there are inherent problems associated with this production. The productive capacity of these lands must be protected and hopefully enhanced to help feed the U.S. population and help the economy. This report addresses the present status and priority research areas to sustain and enhance productivity and protect the resource base of agriculture in the arid and semiarid regions of the United States.

CURRENT SITUATION

Historically there have been significant and, on occasion, fairly rapid shifts in land use in the 17 Western States due to changing economic conditions. During the past 20 years the acreage of cropland cycled from 355 million acres in 1958 to 335 million acres during the 1960's, increasing to 377 million acres by 1977. (Economics, Statistics and Cooperative Service. 1980. Changes in farm production and efficiency. USDA Statistical Bulletin No. 628.) The acreage of irrigated cropland has greatly increased during this period. On a national basis 37 million acres were irrigated in 1958 compared to 58 million in 1977 (USDA, SCS 1977 National Resource Inventories). By far the greatest proportion of the total was in the 17 Western States (48 million acres). Nonirrigated cropland in 1977 in these states totaled 151 million acres . The vast majority of this dryland acreage was in the Great Plains States.

Tremendous advances in crop production technology have occurred in the past 50 years. New technologies greatly reduced on-farm labor requirements while making input suppliers an essential part of present cropland management systems. While the index of all inputs used in agriculture only rose from 101 (1967 = 100) in 1928 to 103 in 1978, non-purchased inputs decreased from an index of 177 in 1928 to 88 in 1978. Over the same period purchased inputs increased from an index of 50 to an index of 119. Farmers shifted from animal power to machine power, animals required less (fuel) feed grown on the farm (non-purchased inputs) and tractors and other machines required off-farm purchased petroleum. In the same way, purchased commercial fertilizers replaced legumes and animal manures, chemicals replaced labor in the form of additional cultivations as weed control measures and hybrid seeds replaced homegrown seed stock. As a result, agriculture became increasingly dependent on agribusiness sectors of the economy.

A rapid increase in productivity accompanied this shift in input structure in U.S. agriculture with the productivity index increasing from 53 in 1928 to 117 in 1978. The productivity increase manifested itself in two ways: (1) to the benefit of consumers, food costs rose by a small amount relative to the cost of other goods and services and (2) an ever-decreasing number of farmers and on-farm workers are required to feed an ever increasing domestic and world population. In addition, the export of surplus agricultural products has increased.

The dependence of agriculture on the non-agricultural sectors has been amply demonstrated by events occurring in the petroleum industry during the 1970's. The cost and availability of petro-chemicals threatens the technological structure on which agriculture has become so dependent.

In many countries of the world food production is not keeping pace with population increase. Through the past decade U.S. agricultural exports, particularly of food and feed grains, have expanded rapidly. The prices associated with continuing high export demand encourage increased production. Production prospects throughout the world are monitored closely and make market prices highly volatile. Export demand is causing conversion of pasture, range and other land to more intensive use. Questions of reliability of future energy supplies and rapidly increasing costs of production cause uncertainty about the future. This uncertainty may inhibit technological advances that are needed and that would otherwise occur.

Price signals which cause farmers to respond by changing resource use often will not reflect the concern of society. Farm operators use resources in response to current and expected near-future price signals. Given the technologies available to them, they make choices on crops, cultural practices, variable input use and land and water use. These resource use patterns result in:

• row cropping land which should be afforded more vegetative protection.

• depleting the stocks of water in underground aquifers so long as available technology, energy, and crop prices favor its exploitation.

• selecting cultural practices which reduce risk or stabilize between-year production levels regardless of the impacts on the soil resource base (e.g., saline seeps or cultivated summer fallow because of less problems with establishing a stand for the subsequent crop, less problems with plant diseases over time, etc.).

Society should have a longer planning horizon. It is the responsibility of scientists studying the use of soil and water resources to develop technologies and to inform the public about the consequences of current land use patterns to current and future generations. In the absence of institutional barriers, near-term price signals serve to allocate resources at a point in time and through a limited planning horizon. Producers respond to price and shift soil and water resource use. Land use shifts between crops and energy prices signal a reduction in use of pumped irrigation water. Decisions about the purchase of durable (depreciable) assets are made on the basis of price and income expectations. But, these price signals are not useful for achieving intertemporal resource allocations over long time periods.

Resource users select practices and crops which minimize economic returns even if these practices use resources in an exploitive manner. Land left fallow and subjected to wind and water erosion loses its productive potential over time. Future prices and resource use values are greatly discounted. The rate of discount increases in the presence of high rates of interest and inflation. Ground water stocks are depleted without adequate attention to future food needs.

The gap between private resource use and public resource use desires will always be present. Soil and water scientists must play an active role in encouraging better stewardship of these resources. New technologies must be developed and adopted to site-specific situations which maintain and preserve our soil and water resources for future generations. With these technologies several options are available:

(1) Incentives as provided through public policies that serve to provide more desirable resource use patterns over long time periods. Such incentives have been used to encourage farmers to establish strip cropping, terraces, etc.

(2) Regulations can be imposed which prohibit certain resource use patterns. For example, practices which may lead to soil losses above certain acceptable levels could be banned by law.

(3) In combination with incentives and/or regulations, educational programs are needed to inform users and the public at large of the consequences of inadequate attention to intertemporal resource allocations. Our potentials to sustain the capacity to provide food abundance for the

U.S. population and continue to export agricultural products to world markets is being threatened. We are exporting soil and water. We must protect these resources so that this capacity can be maintained.

Much progress has been made in the control, management, and prediction of wind and water erosion since the Dust Bowl period in the 1930's. Even so, wind erosion continues to be a major soil conservation problem, particularly in the Great Plains States. In 1977 in the Great Plains States 27,696,000 acres of cropland had wind erosion rates greater than 5 tons/acre-year. This figure, of course, includes the subhumid as well as the semi-arid portion of the Great Plains. Sheet and rill erosion is less significant in the 17 Western States than in the humid east. However, in 1977 there were approximately 20 million acres with sheet and rill erosion rates greater than 5 tons/acre-year. Most of this erosion probably occurs in the subhumid parts of the Western States, but the soil and crop management practices do not differ markedly from the semiarid parts. Several Land Resource Areas in the West have even higher water erosion rates in localized areas that impact on both productivity and downstream water quality. Examples include the Palouse area in the Pacific Northwest and the Blackland Prairie in Texas.

The development of conservation tillage systems, improved farm equipment, and improved herbicides have enabled farmers to reduce soil erosion losses significantly. (Conservation tillage is a system in which surface crop residues are managed to help control wind and water erosion. It is variously called reduced tillage, minimum tillage, limited tillage, no-tillage, stubble mulch tillage, and ecofallow in various parts of the country.) However, world population pressures with attendant demands for increased food production resulted in higher prices which caused large acreages of more erodible, shallow soils to be put in cultivation. Reducing soil losses to tolerable limits on these fragile soils may not be practical with current technology if they are kept in cultivation.

Soil erosion by wind and water has probably reduced crop production potential of lands in both dryland and irrigated areas although the effect of erosion on productivity is not easily estimated. Obviously, it is more critical on shallow than on deeper soils. Improved technology including increased fertilizer use may have masked productivity declines caused by erosion. Still, reduced erosion on much of the cropland in the Great Plains and increased production resulted from new technologies applied in crop and soil management in the last few decades.

Traditionally increased irrigated acreage has been achieved by developing new water sources. Future increases in irrigation water availability are unlikely and, in fact, the total supply is more likely to decrease. In some areas, the quantity of extracted ground water greatly exceeds the recharge rate. The most serious depletion of ground water is in the High Plains areas extending from Nebraska to Texas. Central Arizona and parts of California also depend heavily on ground water. Readily available surface water sources have already been developed and future irrigation water diversion projects will be limited because of cost and public reaction against such projects. Furthermore, agriculture will face increased competitive demands for water from urban, industrial, and energy development entities. Potential water sources which may be exploited more than at present are reuse of urban and industrial wastewater and use of more brackish waters than commonly used. Use of these sources for agricultural production will present unique management demands and require that the water be used efficiently.

Leaching in excess of that required to reduce salinity in the root zone to the desired level contributes to water inefficiency. Leachate that replenishes ground water may not be considered as a resource loss, but it may lead to inefficiency because of pumping costs to return the water to land surfaces. Furthermore, nutrients and pesticides are often leached and contribute to water degradation and to a resource loss in the case of nutrients. The evaporation part of evapotranspiration is inefficient water use. Runoff and conveyance losses which are not captured and used somewhere in the system also contribute to inefficiency.

Increased water use efficiency (unit of crop yield per unit of water used) can be achieved by changes in irrigation management. Surface irrigation is generally less efficient than sprinkler or drip irrigation. Variability in soil properties across a field allows variable infiltration and percolation rates. Water is generally applied to supply sufficient water for the crop in zones of least water infiltration thus allowing excessive leaching in other parts of the field. Sprinkler and drip systems can minimize soil variability as a factor in leaching by controlling the rate of water infiltration, rather than allowing the soil to control the rate. Dead-level surface irrigation is a recent advance toward attempting to increase efficiency of surface applied water. Drip irrigation by limiting the wetted soil surface area can often increase water use efficiency by reducing the evaporation component of evapotranspiration .

Although shifts in irrigation systems can lead to increased water use efficiency, these shifts are not always economically feasible. Presently water is generally inexpensive and the cost for saving the water exceeds the value of the water. However, negative effects to other people of water degradation by excessive leaching are not accounted for. In the future, however, competing demands for water and increased pumping costs will make water more costly and provide incentives for water conservation.

Shifts in irrigation practices will require shifts in other farm management practices such as fertilization, weed control, insect control and plant disease control. Research is required to provide information on effective integrated water-nutrient-pest control management systems. From experience, we know that shifts in one management practice requires shifts in other management practices. The scientific community must be prepared to provide information for intelligent shifts in farm management.

Also, it is important to initiate research to provide information about the consequences of government programs on resource use and conservation. Some programs are designed to control supply while protecting and stabilizing farmer income; other programs are designed to promote resource conservation. Often, objectives of supply control and resource conservation are carried in the same act; but, the two objectives may be competitive rather than complementary. Some examples include programs to stabilize incomes in high risk farming areas tend to encourage cropping of land which should remain in permanent cover. Acreage allotments may have led to excessive amounts of summer fallow which unduly exposes soil to wind and water erosion; or soils which should be managed with a legume rotation were placed under continuous row crop production in efforts to keep the commodity base acreage as great as possible. In much the same way specifically designed resource conservation programs work counter to supply control programs by stimulating output expansion under the label of resource conservation.

As research needs are considered, the marked contrasts of soil characteristics and climate stresses of the western arid and semiarid regions compared to the humid areas of the East and South must be clearly recognized. Management practices that maximize the conservation of as much of the limited rainfall as possible and its efficient use by crops address the primary limiting factor of the area, a problem enhanced by the high evaporative demand of the area's stressful climate. Humid areas must deal seasonally with problems of excess water and drainage. In semiarid areas the relatively unleached soils contain large quantities of nutrient elements such as phosphorus and potassium through the entire profile in contrast to the leached subsoils of humid areas. Some elements such as phosphorus may be in less plant-available forms, but the possibility exists to breed crop varieties that can use more of such forms, thus opening up a vast quantity of plant nutrients. Research is needed to improve fertilizer use efficiency through different forms, rates, methods, and time of application. Also needed are recommendations based on soil tests. There is a great need for added research on conservation crop production systems for the semiarid regions to improve efficiency in using the limited precipitation, increasing stability of crop production in the stressful climate, and in capitalizing on the nutrient-supplying capacity of the soils.

In the northern Great Plains of the U.S. and Canada, approximately 2 million acres of cultivated drylands are affected by a salinity problem caused by excess water penetrating below the root zone during fallow or idle periods between crops. This water moves through salt-laden substrata to impermeable or slowly permeable layers and forms a perched saline water table. It then flows laterally downslope from the recharge area, resurfaces at a lower elevation, evaporates, and deposits its salt load. Under native sod, little water entered the substrata because precipitation-evapotranspiration cycled almost entirely within the root zone. Current technology, emulating this process, indicates that when soil water is adequate, a crop should be seeded rather than traditionally summerfallowed every other year. However, the technology is not fully utilized.

The burning of fossil fuels has increased and will continue to increase the carbon dioxide concentration in the atmosphere. The weather and climatic effects of increased atmospheric carbon dioxide are unknown, but some scientists predict increased variability in weather in the Great Plains. The predicted increased global temperature associated with increased atmospheric carbon dioxide indicates an increased vulnerability to our crop productivity and food supply.

One negative effect of the farm-to-urban migration is the greatly reduced number of people with farm experience in the younger generations available to fill jobs in agriculture. Many students in colleges of agriculture today have had no practical farm experience. Meeting the trained personnel needs of the highly technological agriculture that is evolving presents educational institutions with new and difficult challenges.

The problems associated with conservation technology on agricultural lands in the semiarid and arid regions will be presented in the following discussion. Associated research needs to improve conservation and production technologies will also be presented.

STATEMENT OF THE PROBLEMS

Dryland Agriculture

Crop yields in the semiarid region are low in most years due to water deficit at some time during the growing season. Precipitation is characteristically low and erratic in semiarid regions and, thus, limiting for dependable crop production. Evaporative demand is high. Bare soil surfaces, mechanical compaction, lack of crop residues, low water intake rates, shallow soils, weed growth, and other factors all limit infiltration and soil moisture storage for crop use. Regional instability of production results in undesirable economic consequences both to the growers and to the nation's economy. However, with improved technology developed through research, both the amount and stability of crop production can be markedly increased while protecting the land resource base.

Direct evaporation of water from soil is a natural phenomenon and is part of the energy balance of the earth. Water evaporated directly from the soil cools the environment but does not pass through the plant and thus does not contribute to crop production. Evaporation from the soil also brings salts to the surface which may cause germination or infiltration problems, particularly if sodium is present in damaging amounts. The amount of water lost by evaporation commonly accounts for up to one half of the total evapotranspiration.

Crop residues are the most universally available material used to reduce evaporation at the soil surface. Increased knowledge concerning the management of crop residues could reduce evaporation losses and increase soil water storage. Surface residues reduce evaporation by shading the soil, insulating against energy conduction, and by lengthening the water vapor diffusion path length at the soil surface. They also absorb the impact of raindrops, help maintain higher infiltration rates, and slow runoff. One main effect is to keep the surface layer more moist during the drying process so that less rain is needed to rewet the soil to a given depth, and thus a given quantity of water penetrates deeper where it is less subject to evaporation losses.

In dryland areas the amount of crop residue available is often less than that needed for either erosion control or water conservation. Each tillage operation reduces the quantity of remaining residues. Substitution of herbicides for tillage in controlling weeds shows promise to conserve surface residues, but much more research is needed. Other innovations are needed to improve the role of crop residues in conserving soil and water.

As geneticists, agronomists, and plant physiologists have developed higher yielding, pest resistant crop varieties, they have simultaneously improved water use efficiency. Conservation tillage systems which maintain more surface crop residue greatly modify the soil environment from that of conventional systems under which most existing varieties were developed. Much breeding effort is needed to develop varieties for these altered environments. Additional progress can be made in breeding specifically for resistance to water and temperature stresses, to plant diseases and insects, to nutrient imbalances, and to climatic constraints such as photoperiod and vernalization. The goal should be to determine and understand the physiological basis for potential yield increases so that physiological-genetic variations can be used in breeding higher yielding varieties. Stress symptoms in plants is an expression that damage has already occurred. Consequently symptoms cannot be used to indicate when soil management is required to alleviate stress. Research is needed to utilize weather, soil, and plant information to predict, on a stochastic basis, the dynamics and severity of field environmental conditions likely to produce plant stress. When linked to climate these predictions also address the occurrence of risk associated with known soil and plant combinations. With such a systematic approach to predicting stress, research results on genotype resistance and resistance mechanisms can be more quickly applied.

Current tillage and cultural practices are often inadequate for creating and maintaining soil physical conditions suitable for plant growth. There is a need to develop meaningful methods for assessing soil water, soil temperature, soil structure, and other physical factors on a field basis and to determine how soil and crop management practices affect these regimes and their impact on crop growth. Excessive or improper tillage practices as now employed may be causing compacted zones that interfere with root growth and water infiltration. The present inability to characterize soil physical conditions critical for plant growth and then to create and sustain them throughout the growing season limits both a definition of suitable tillage objectives and the design and use of equipment or cropping sequences to achieve them. Research is needed to develop economic methods for modification of soil physical properties, especially reduction of compacted zones which may reduce crop productivity.

Soil water, soil nutrient status, soil oxygen, soil structure, tillage practices, and soil amendments affect photosynthetic efficiency of plants directly or indirectly. There is a need to develop and utilize integrated production strategies which would include better use of soil and water resources and which would maintain or improve production, reduce production costs, control erosion, and avoid production hazards. The concurrent development of field research and mathematical models is needed to study and simulate interactive effects between soil, plant, water, nutrient, and atmospheric conditions under management methods designed for a) conserving energy, soil, and water, b) creating an optimum soil-plant-atmosphere environment, c) and increasing crop production.

Currently we do not have systematic methods to ascertain grower risk when he adapts conservation tillage methods. A systematic and quantitative method of analyzing these risks would help to determine critical deficiencies in proposed conservation tillages and research approaches to develop new tillage techniques. This process would also bring into account climatic and soil impacts on risks and identify other factors of importance, such as plant performance and hazard for disease, weed, and insect pests.

Research is needed in the selection and utilization of legumes for dryland cropping systems to provide soil protection and decrease the need for nitrogen fertilizer. Efficient crop residue and organic nutrient (biological-N) management systems coupled with the application and conservation of inorganic nutrient supplies have not been developed for conservation tillage cropping systems in the semiarid region. Modeling of nutrient cycles, particularly N and P, has provided some insight and added to our understanding of the soil-plant-environment factors affecting recycling rates, but the interaction of soil and crop management can only be investigated in well-designed field experiments. Such experiments must address water conservation, soil erodibility, crop residue management, soil fertility, etc., in a combined rather than in a segmented manner when investigating the soil and plant environments needed to optimize the production of each crop within a cropping system. Such dryland cropping systems must be site specific across the great diversity of soil types and climatic circumstances of the semiarid region.

The present trend of converting marginal lands, formerly in grass, to cropland because of high grain prices may accelerate in the future if economic incentives become even more favorable. Limited and erratic rainfall increases the economic incentive needed and also increases potential wind and water erosion. New technologies such as conservation tillage will have to be used, more drought tolerant crop varieties developed, and integrated management systems devised to successfully and safely farm these lands. There is a need to develop flexible crop sequencing systems that match crop water requirements with existing soil water at planting time and probabilistic anticipated growing season precipitation. Inherent in this concept of sequencing different crops are soil water recharge during periods between crops and strategies to interrupt disease cycles and crop-specific weed problems. The improved soil and crop management systems will need improved weed, insect, and disease control methodology if crop production is to be increased and/or stabilized. Because of increased pressures to utilize biomass and/or crop residues as energy sources, careful evaluation of the potential short and long term effects on soil erodibility and crop production must be considered. Engineering practices and structures for improved water management and conservation should be an integral part of the management system. These might include water harvesting and snow management schemes for improved crop production under water short climatic conditions, terracing, furrow damming to reduce runoff and conserve water, and surface soil modification to increase intake rates for deeper storage of water.

Biological control of soil borne insect and disease pests can be exercised more often and systematically in croplands if better information is obtained regarding the effects of soil water content and water potential on organisms that reside in the soil. Antagonism between various microorganisms and pathogens (or insects) depend on field soil water regimes. Advances require coordinated multidisciplinary inputs and must more carefully describe and evaluate soil water, soil temperature, soil aeration, and plant residue disposition.

The stressful climate in the 17 Western States with low, erratic rainfall and great diversity among the soils involved are factors that make achieving reliable weed control with herbicides without residual carryover quite difficult. Fields with large variations in soil pH, soil organic matter, and soil texture complicate herbicide management to achieve reliable weed control. Soil, crop, and weed scientists need to address these problems in systems approaches.

Achieving profitable and stable crop production along with effective conservation of soil and water in arid and semiarid regions requires that farmers use conservation production systems which effectively integrate a wide range of practices and factors such as choice of crops, varieties, cropping sequences, land farming, wind barriers, shelterbelts, terraces, strip cropping, contouring, pest control, and irrigation water management. This integration process is complex. Research is needed to characterize the important interactions that must be considered in combining the various technologies into the most workable systems.

Wind and water erosion continue to be a problem on dryland. The importance of the problem is difficult to assess because of the erratic occurrence of severe storms, accumulative effect over the span of cultivation, and inability to relate crop production, economic cost, pollution, etc., to soil erosion (and erosion control systems). Fluctuating prices have caused many unwise land conversions which have led to cropping of marginal land in marginal climates that have caused erosion by wind and water. Rehabilitating these lands once cultivated will be a formidable task.

Conservation tillage is a very promising method for controlling soil erosion, runoff, and water pollution in the future. There is likely a complementary relationship between conservation tillage and soil, water, and energy conservation. Systems need to be developed that are capable of controlling wind and water erosion and producing the desired combination of roughness, surface residue cover, and seedbed conditions for the wide range of soil and climatic conditions encountered in these regions. The systems should satisfy the differing seedbed requirements of individual crop species. Use of herbicides as an alternative to some tillage operations merits special research attention. The fertility requirements of conservation tillage systems as well as tillage interactions with weed control, crop sequence, diseases, insects, and water conservation need to be identified.

In order to provide long-term protection of the soil resource base, Soil Loss Tolerance Values are needed that will allow accurate assessment of the long-term effect of soil erosion on soil productivity. Such an assessment will require a knowledge of the effect of soil erosion on the various soil-related factors affecting crop yield and the determination of their net effect under different crop, soil, climatic conditions, and management systems. The economic impact of soil erosion on crop yields needs to be accurately assessed.

While there is universal agreement that soil erosion depletes soil productivity, the relationship between cumulative soil loss and productivity is not well defined. Until this relationship is adequately developed and verified experimentally, selection of management strategies to maintain crop production and to control erosion will be impossible. Lowering of soil productivity is often masked by improvements in soil and crop management, improved plant varieties, and increased use of technology. Nonlinearity of the soil erosion process compounds the difficulty of detecting productivity losses.

The best approach to solution of the problem will probably be a combination of mathematical modeling and field experimentation. Objectives should include (a) developing a physically based model to simulate the processes (hydrology, nutrient cycling, plant growth, soil properties, crop management, including weed, insect, and disease control, soil management including tillage, and environmental quality) that affect soil erosion and soil productivity, (b) model verification by field experimentation in critically eroding areas, and (c) providing model outputs necessary for economic analysis concerning the cost of eroded soil. Successful completion of this research would assure maintenance of long-term soil productivity while optimizing crop yields through development of improved erosion control and soil management systems.

Soil Loss Tolerance (T) values have been used extensively in soil conservation work and are increasingly being used to achieve a broad set of social goals, including nonpoint water pollution control and reduction of downstream damages from sediment. However, on some soils T-values may be too liberal for long-term productivity maintenance, while on other soils, the current levels may be unnecessarily restrictive. Different T-values may be required to meet the different objectives desired. Thus, both the criteria used to develop T-values and the objectives desired must be clearly defined. Regional soil, environmental, and crop production factors will need to be considered.

The greatly expanded use of the Universal Soil Loss Equation (USLE) in areas outside of the geographical location and climatic area for which it was developed and for other purposes than originally developed , such as by state agencies for regulatory purposes, necessitates the refinement of guidelines for the use of the USLE and for determining the limits of its applicability. A new data base and possible modifications may be required for situations where the USLE is currently limited in applicability. In the Pacific Northwest, for instance, there is a need to account for severe runoff and erosion that may occur on frozen soils and to develop appropriate soil management for control. Modifications or improvements are needed also to better account for erosion during single storm events or account for soil detachment transport and deposition. Ultimately, physically based equations are needed to refine soil management guidelines for runoff and erosion control.

The climatic factor of the wind erosion equation was developed for the semiarid Great Plains and is not well-defined in the more arid West. It needs modifying to more realistically account for surface soil dryness and to exclude nonerosive winds. Also, soil erodibility, being influenced by soil, climate, and management is extremely dynamic. Improved guidelines are needed for relating soil erodibility to those variables influencing erodibility in the arid and semiarid environment.

Irrigated Agriculture

Some of the problems of dryland agriculture also exist in irrigated agriculture, but the greatest emerging problems involve diminishing irrigation water supplies, decreasing water quality due to increased salinity, and rapidly increasing water costs due to increasing energy costs and pumping lifts. Competition for limited water supplies and water costs will continue to increase in the future. Water quality will continue to decrease with increased use. These problems, along with others listed below, will require major new concepts in planning for multiple water use, new and improved technology for storing, delivering and uniformly distributing water on irrigated lands, and improved management technology to increase production per unit of water consumed. Management technology affects the entire water supply and delivery systems, conjunctive use of surface and ground water storage, farm irrigation systems, and the timing and control of the amount of water applied. In some areas it also includes the management of drainage systems to control the water table level below the crop root zone, the collection, reuse and/or disposal of moderately saline drainage water, and the variation in cropland irrigated depending on available water supplies for the season. Some of the specific research needs for specific problems are described below.

All irrigation systems apply water to the soil at unequal amounts from place to place. Surface irrigation methods may result in poor uniformity if not properly managed. Better methods are needed to apply the water to the soil where and when needed. New design approaches for these systems are needed to modernize existing systems. Irrigation system designs are needed which more properly reflect the soil properties and crops grown. New sytems such as trickle and "surge flow" furrow irrigation may require site specific management practices but have potential for high efficiency. Limited tillage methods need to be developed that are compatible with different irrigation systems. Better management strategies are needed to modify infiltration capacities.

A pressing need is to be able to prescribe, in reasonable terms, which irrigation system should be used in a given site and cropping system. Land is going into sprinkler systems where surface systems may work as well and be more energy conservative with less capital. Sprinklers may be used because they are so accessible and convenient, even though they are energy intensive. Various water application methods need to be evaluated in terms of capital costs, longevity, energy demands, application efficiency, application uniformity, availability of commercial components, degree of automation, soil and crop characteristics, etc. New systems need development, i.e., the "surge" flow furrow system with required automation.

The increased energy costs have emphasized the advantage of low pressure sprinkler systems, but these may require better soil management practices to minimize runoff and erosion. Management of amount and timing of water application needs to be integrated with soil properties so that water can be applied uniformly without erosion or other deleterious effects.

Increased energy costs have caused a substantial increase in fertilizer costs. Consequently, more careful use of fertilizer and better irrigation management are essential in the future. In many irrigated areas, over-irrigation and over-fertilization are the prevailing practices. This may cause problems such as high nitrate content in irrigation return flow and ground water. Better understanding of nitrogen interactions and movement within the soil would be desirable but is a difficult problem to solve. Much could be done with present technology if accounts were kept on the water and fertility balance of the entire system.

Especially with nitrogen, ways to increase fertilizer use efficiency (nutrient in crop/nutrient applied) in combination with water use efficiency are needed. Such practices as foliar feeding, applying fertilizer in the irrigation water, and fertilizing according to plant demand curves, need to be evaluated better. A better understanding of the interactions among irrigation, plant nutrient needs, and conservation tillage systems, especially over a period of years is also an important consideration.

Increasing energy costs and decreasing availability of irrigation water will emphasize water use efficiency. Where there is very little rainfall, production per unit of water applied is greatest if water supply just balances plant needs. This necessitates better knowledge of plant water requirements. Water stored in the soil could be used more efficiently if it were measured. Even consideration by farmers of stored soil water would improve overall irrigation efficiency. Knowledge of water balance techniques could be applied with great benefit.

Where irrigation supplements substantial rainfall, the goal should be to use limited irrigation water to increase precipitation efficiency. In many areas irrigation water is not available throughout the season or water supply is insufficient for the land available. The management decision of when to apply the water and, to what land and crop, are not simple. Computer "games" could simulate answers to many of these "what if" questions. Field knowledge and common sense must be exercised to be realistic. More research is needed on the relation between rainfall probabilities and supplemental irrigation.

Irrigation water, appropriated and used by municipal-industrial users can often be advantageously reused for irrigation. Irrigation can often dispose of polluted water and minimize some potential pollution problem or difficulty. An example is the use of the waste water from electrical power plants or waste water from food processing for irrigation. In other cases the soil may be used as a filter for pollutants in food processing wastes. Each situation has its own peculiar problems but many common management practices can be used to advantage. Research opportunitites may be available to develop understanding of processes and how they can be solved or tolerated by agricultural users. Financial resources are available for this shared use of water so an opportunity for development of practical dual use systems exists. However, government regulations often make the shared use of water by industry and agriculture very difficult to accomplish.

Institutional constraints built into existing water rights laws may limit economic incentives to increase water-use efficiency. The maintenance of water rights is of more importance to the farmer than efficiency. We need ways to protect historic water rights while eliminating inefficient water use. Increasing water prices could be a possible way to achieve this goal.

Future irrigation problems with respect to salinity will undoubtedly increase. Irrigation supplies will be both more saline and limited in amount. Limited or reduced leaching needs to be tested for a wide variety of situations for practical applicability. More attention should be paid to management systems that permit salt to accumulate in the soil for some time to be removed by periodic heavy rainfall or short periods of leaching. This will allow for better control of leaching and possibly allow the drainage water outflow at a better time.

Many salinity management schemes have been shortsighted in the past because research results from soil columns or from small field plots have been extrapolated to large areas without considering factors introduced by the large system. The hazards of partial soil salinization using reduced leaching with present irrigation systems such as furrow are enhanced. Increased water use efficiency with associated reduced leaching has been proposed for salinity improvement without considering alternate uses for the water saved. The implication is that the water saved will go downstream without any additional increase in salinity. However, the farmer sees this saved water as an additional source of water which he can use to irrigate more acres which increases evapotranspiration and salt concentration. Such practices may actually decrease return flow and reduce water available to downstream farmers or alternate users.

There is a need for better methods to routinely monitor soil water content before irrigation. Instrumentation is needed to quickly measure soil water over a large area. Present methods are point measurements that require expensive equipment and many measurements to characterize field soil water content. In many instances, this need could partially be met by using simple computer models that keep track of the soil water balance, perhaps with periodic remotely sensed inputs. However, these models require better estimates of evapotranspiration, rooting depth, and available soil water.

Better methods are also needed to evaluate the existing soil salinity profile in the field. While these data are not needed as frequently as soil water content, the same problems of large volume sampling and spatial variability exist.

Crop selection can be used to conserve water and energy. Consumptive use requirements of some crops for water correspond more closely to the natural environment. For example, in the Central Plains winter wheat can be successfully grown under irrigation with some fall and early spring applications of irrigation water. Providing water during the high evapotranspiration months of July and August is avoided. But breeding and selection of winter wheat varieties which are responsive to irrigated conditions are needed.

Crops also vary in the nature of their response functions to limited water applications. Grain sorghum has been treated as a crop which produces a relatively large response to limited amounts of irrigation water. Added breeding to enhance this trait is needed. Research has also shown the importance of timing of irrigation water applications. The scheduling of irrigation applications to coincide with critical stages of plant growth will serve to promote both energy and water use efficiency.

Cropping systems need to consider complete energy managements. This includes not only tillage, irrigation scheduling, crop sequences (including legumes), herbicides, insecticides, and fertilizers but crops that are a direct source of energy, i.e., oilseed crops. These crops could provide an alternative source of fuel for such on-farm activities as drying, heating, cooling, irrigating, and cultivating. This would provide the operator with additional flexibility in management of his harvested crop depending upon the existing economic situation. However, certain energy production potentials also threaten to aggravate soil erosion; the removal of crop aftermath to serve as biomass for energy production can have deleterious effects on the soil. Alternatively growing specialized biomass crops which call for the removal of the entire plant must be constrained by the capacity of the soil to sustain such practices.

Water use efficiency is increased by selection of improved varieties of proper crops for production at the level of irrigation possible. The case for additional breeding effort to serve the special site needs under irrigation in the semiarid region is as strong as for the nonirrigated situation.

Effective weed control is essential in all successful crop production systems. Use of herbicides is a rapidly developing and changing technology that must be integrated into irrigated crop production systems along with tillage, fertilizer, and water application procedures. Soil characteristics have important influences on the activity and persistence of different herbicides.

Basic research on the physiological responses of various crop species and varieties to limited irrigation under a wide range of soil and climatic conditions is needed. When can water supplied be reduced without serious reduction in yield or quality of the marketed product? This applies to modern high-frequency irrigation systems (solid-set and center-pivot) as well as to conventional surface or trickle methods. A similar problem of increasing importance is how to maximize yields per unit of water with a limited supply of water to spread over a given area.

Most wind erosion problems on irrigated lands occur during the nonirrigation season allowing the use of the same management techniques for irrigated land as for dryland. Often under irrigated conditions, there are larger amounts of residue, which, when properly managed, can effectively control wind and water erosion. Many cash crops grown under irrigation leave little residue and necessitate control practices other than residue cover.

Water-caused soil erosion on irrigated land is a serious problem particularly with furrow irrigation systems on sloping lands. Land leveling has potential to reduce erosion. A need exists to develop methods for stabilizing soil for these systems. Use of limited tillage with large amounts of surface residue presents problems with furrow irrigation because water flow is retarded. On some clay soils, slower flow is beneficial to allow deep penetration of water. The possibility of developing management systems to minimize soil erosion with systems such as "surge flow", etc. should be studied. Treatment of the soil to stabilize aggregates has been studied for years as an academic exercise. Present and future economics may provide the stimulus for large scale field treatments.

RESEARCH PRIORITIES

Water is the main factor limiting plant growth in the semi-arid and arid regions of the country. The ability of soils of the region to sustain crop and animal production is declining because of erosion, salinity, and nutrient loss. Crop yields usually do not reflect these adverse factors because of increased technological inputs.

New conservation-tillage practices show promise for decreasing the rate of soil erosion. However, these practices must be integrated into systems that optimize crop production and use soil water efficiently.

Irrigation has stabilized production in much of the semi-arid and arid regions. Most of the high-value fruit, vegetable, and other crops that require warm, dry climates are irrigated. Competition for limited surface and ground water supplies has resulted from expanded irrigation as well as from increased water demand by cities, industries, and others. Ground water depletion, together with increasing energy costs, is causing some irrigated lands to revert to dryland agriculture. Water quality will inevitably be degraded as use and reuse of water intensifies.

Much previous soil and water management has been directed toward eliminating crop stress by irrigation or fertilization, or both. Elimination of all water stress causes excessive irrigation which, in turn, may lead to the leaching of plant nutrients and pollution of ground water. In irrigated areas, a better approach for increased conservation of soil, water, and energy may be to use limited irrigation and adapt crops to withstand the stress. In dryland areas, by knowing soil moisture and the probability of seasonal precipitation, one could select the optimum planting time, crop variety, and fertilization level.

Within-season monitoring of crop condition by remote sensing can provide information on the extent and severity of stresses such as drought, disease, insects, and fertility. To use this information effectively, the agricultural research and extension communities must help determine the specifications for an agricultural remote sensing system.

Research needs related to sustaining the land and water resources on our arid and semi-arid land include:

1. Conservation Tillage to Protect Soil and Water Resources

Research is needed to further develop workable and acceptable tillage practices that utilize surface crop residues to protect the soil from wind and water erosion. Emphasis should be given to water conservation, weed control, fertilizer requirements, and effects of pesticides on environmental quality. The full benefit of conservation tillage for conserving soil and water will be realized only if production is improved or economic incentives are provided.

2. Efficient Use of Diminishing Irrigation Water Supplies

New technology, compatible with changing economic and institutional arrangements, is needed to increase water-use efficiency and to augment water supplies by more effective use of brackish and other poor-quality waters. Additional knowledge is needed to effectively control and reuse moderately saline return flows and to enhance precipitation of neutral salts in subsoil. Also needed are improved methods to control or prevent nutrients, sediment, and pesticides from entering return flows. Impacts of short- and long-term water reallocations on food and fiber production need to be predicted.

3. *Conservation Production Systems for Improved Farm Management*

Research is needed to integrate a wide range of climatic, biological, mechanical, and management practices into systems for farmer use that meet the basic objective of profitable and stable crop production along with conservation of soil and water resources. Practices comprising such systems include choice of crops and varieties, cropping sequences, conservation tillage, land forming, wind barriers, terraces, strip cropping, contouring, fertilizer use, pest control, and irrigation water management.

4. *New Technology to Cope With Production in Stressed Environments*

Research is needed to combine weather, plant, and soil information to reliably forecast the dynamics and severity of stressed field environments and associated crop responses. Increased effort should be made to develop remote sensing for assessing and predicting soil, water, and crop conditions. Techniques are needed to identify the stress-resistance mechanisms and cultural practices that can be used to manage soil and water resources to increase yields. We need to determine the proper times to apply limited amounts of irrigation to optimize water-use efficiency.

5

Stephen C. Smith, Chairman

David Allee, Sandra S. Batie, A. C. Basu,
Charles Benbrook, John Carlson,
Emery N. Castle, Homer Folks, Albert R. Grable,
Roy M. Gray, Henry Hart, Earl Heady,
James Hite, Lawrence W. Libby,
Gordon Rausser, Gerald B. Welsh

POLICY

Alternatives and Consequences of Conservation Technology on Agricultural Land

As the nation enters the 1980s, soil and water conservation is moving into the center of the policy arena. The effectiveness of present policies seems to have plateaued. They are broadly recognized as not meeting present problems. In 1977, soil erosion exceeded 5 tons per acre per year on 23.3 percent of the nation's 406.3 million acres of cropland (Miller, 1970). Report upon report concludes that soil losses are serious enough to reduce agricultural production in the future. Other important soil and water conservation- related problems include nonpoint-source pollution and water quality and quantity issues.

Most soil and water conservation programs began in the 1930s. Story has it that dust from the Great Plains arrived in Washington, D.C., while Hugh Bennett was testifying for the creation of the Soil Erosion Service. Eastern as well as Western congressmen experienced first-hand the reality of wind erosion as soil particles settled on the nation's capitol.

Soil erosion from water was also real. Perceptive commentators on the American agricultural scene in the nation's earliest days decried agricultural practices with incessant loss of soil accompanied by locally declining productivity. Southern cotton agriculture moved westward for a variety of reasons, leaving eroded land to succeeding generations.

Serious soil erosion occurred in the Palouse before the turn of the century and escalated to the point that 10 percent of the rich cropland had lost its original topsoil by World War II. Between 1940 and 1960, the average estimated loss per year of topsoil continued at 8.8 million tons.

Such events throughout the nation led to soil erosion becoming a priority policy issue in the 1930s. National programs were created that (in modified form) remain in place today. These include technical assistance to local conservation districts through the Soil Conservation Service, federal grants-in-aid to assist farmers with paying for a share of the cost of conservation practices so that off-site beneficiaries bear part of the burden, and regional programs, such as the Tennessee Valley Authority and the Great Plains Conservation Program.

In the intervening 50 years, agriculture made phenomenal technical and organizational adjustments. Total and unit production increased dramatically. The incentive to maximize current net income carried the day. Policies and institutional structures promoting soil erosion control and water conservation were not strong enough to achieve these goals. The goals appeared counter to the pervasive economic and social forces and resulted in continued high rates of soil depletion.

Conservation practices were applied, but not widely enough. Cost-sharing too often missed the target of putting into place the highest priority-conservation practices. Funding was inadequate (a) to install enough high-cost practices, particularly as credit and the capitalization of the farm business changed, and (b) to be distributed to enough owner/operators to achieve greater conservation. And farm situations across the country varied widely. Broad national policies were not tailored to each situation.

Notwithstanding these significant inadequacies, soil-conserving practices were put into place and favorable results can be cited in many areas. The rate of soil depletion today is less than it might have been. But the limited gains of the past are now vulnerable to new conditions. The tide of changing farm organization — larger but fewer farms(some call them super farms); increased size of equipment(equipment today is mammoth in comparison to yesterday); increased capital requirements; high interest rates; reduced on-farm production of feed for workstock, including hay and pasture; increased confinement of livestock; etc. — have raised the pressure for high levels of cash flow, resulting in fence-row-to-fence-row cultivation. Depletion, not conservation, is widespread.

Policy performance, however, has not been in total disarray. Individuals and communities can be cited that are exceptions to the general pattern. Many soil- and water-conserving practices are in use. A gnawing feeling, however, is widely held that we should conserve the soil and use our water wisely if it were only profitable today—for me.

The difficulty of implementing a successful soil and water conservation policy has not gone unnoticed by analysts and research scientists. New soil-conserving practices have been developed to fit today's conditions. A foundation of economic and sociological research has been laid. Also, experimental policies have been implemented by states and local communities as federal dollars have become limited. States have passed innovative regulations, and the practice of state and local cost-sharing has been tried under a variety of conditions. While cross compliance has been debated at the national level, some local units of government have used it when spending their own money. The variety of national, state, and local policies provide a body of experience waiting to be analyzed.

The purpose of this panel is to assess, as of 1981, the 50-year stream of events that shaped soil and water conservation policy and to provide a context for developing future policy-relevant research. Our intent is not to explore the technical refinement of research, although this is essential, but to make the point that future investment in policy relevant research will improve tomorrow's policy deliberations. Misconceptions about what is possible will continue to frustrate policy agreement in the absence of such an undertaking.

The focus of soil and water conservation is the rate of resource use—rates that extend use into the future are more conserving than rates that use the resource today. In addition, conservation policy is concerned with who receives the benefits, with who pays the cost, with when the benefits and costs accrue, and with what motivates a farm operator to take conserving actions. Answers to these questions must be sought in terms of the political, social, and technical milieu. The panel's objective is to categorize the areas of research on these topics that are relevant to tomorrow's policy and hence, deserve funding.

As noted, the present structure of conservation policy institutions had its inceptions in the 1930s. To understand today's research needs, we will note this structure and the forces that shaped it. The following points partially characterized the situation 50 years ago:

1. The great depression
2. The great drought
3. Serious soil erosion and lack of water conservation
4. Dominance of animal power with little electricity and scientific agriculture
5. Farm markets—many buyers and sellers
6. A politically powerful farm block in Congress as well as farm interest control of legislatures
7. A recognition that the individual land manager had to take action
8. Awareness of benefits off-site and in the future, thus legitimizing the expenditure of public funds

THE GREAT DEPRESSION

The great depression was an overpowering influence and brought forth many governmental programs with a problem focus, but also with the clear intent "to get money to the people." The Civilian Conservation Corps and other such programs were intended to combat erosion and improve water management while giving desperately needed jobs and job training to people. Conservation payments of the 1930s were to yield a conservation benefit, but they were also to bolster farm income and reduce the supply of farm products on the market.

The rationale for spending money during the great depression does not exist in 1981. Today, inflation is the problem, with high prices and high interest rates, plus selected unemployment and the need for industrial and capital restructuring. The watchword is a reduction in governmental expenditures to combat inflation. Government payments and jobs are not seen by many as the appropriate answer. Conservation practices need to be devised that will minimize public payment to achieve adoption. Thus, research must continue into the economics of farming systems that can maintain conserving practices and not drop income and cash flow below acceptable levels.

During the past 10 years several models have been developed that explore these issues; however, this remains a priority research area. Possible alternative farming systems need exploring for the wide variety of conditions and types of farming throughout the country. The development of basic models adapted to individual farm planning with alternative soil and water management systems programmed for home computer use is also an area for future development.

THE GREAT DROUGHT

The great drought of the mid-1930s created a national sense of resource crisis that reinforced the economic calamity. Clouds of dust blew across the continent, evidencing the vulnerability of the very basis of the nation's greatest industry, agriculture, and of its most traditional means of livelihood. Fifty years later, floods and droughts still occur. But it is the energy crisis that periodically galvanize national attention to the extent the Dust Bowl once did.

To be sure, careful inventories and accumulating scientific evidence, as mobilized in the Soil and Water Resources Conservation Act (RCA), show that our soil is now eroding at dangerously high rates. Scientific attitude surveys (for example, Louis Harris and Associates, 1979) show that a strong majority of the people recognizes the problem and prefers government action, even at some cost. But there are at least two sharp differences between this public orientation and that of the 1930s. First, a number of crises confront us in the 1980s, among which the threat to the soil is neither the most dramatic nor the most urgent. Second, soil conservation has been addressed for decades by well-known federal programs and federal-state-local organizational structures. If people are to rally for a great new effort, they will want to know how it improves upon the old one.

SOIL EROSION/WATER CONSERVATION PROBLEMS

Both the 1930s and the 1980s documented the seriousness of soil erosion and water conservation as a public problem. This does not need repeating. Although there are many differences in the character of farming between the two periods, the underlying cause for soil erosion and lack of water conservation is the same—conserving farm practices and organization were not profitable enough to incorporate into farming systems on a widespread basis. Both periods have a similarity in this respect.

TECHNOLOGICAL CHANGE

Agriculture in the 1930s was dominated by animal power, with accompanying farm organization and practices (for example, on farm production of feed and pasture for workstock). Electricity was just emerging into widespread use. The structure to implement conservation policy was based upon this model of farming. Furthermore, the impact the agricultural sciences was just beginning to be applied in such a way as to result in very high rates of productivity. Hybrid corn, for example, was not in widespread use.

In the intervening years, agriculture has been transformed with both new farm practices, on-farm organization, and much better marketing systems. The farms of the 1980s are a marked contrast to those of the 1930s; and according to some, many firms producing agricultural products today are not recognizable as farms. In fact, a current public debate is raging: Is the production of an agricultural product still farming when certain thresholds of size, concentration of livestock, capital intensity, specialization, decisionmaking structure, etc. are passed? When does a dairy farm cease to be a farm—50 head, 100 head, 500 head, 1,000 head, etc.? When does the production of eggs cease to be farming–250 birds, 2,500; 250,000; 2,500,000, or larger? When does the production of any agricultural product cease to be farming due to a change in the organization of the activity? Do the larger food-producing firms of the 1980s have the motivation of the smaller family farmers of the 1930s to save the soil asset so it may be passed on to the next generation—stockholders or a son, daughter, or in-law?

As soil and water conservation programs are shaped for the future, the transition from low-cost oil and electricity to high-cost power, with possible supply disruption, must be taken into account. High cost energy and shifting sources of energy will affect the impact of new scientific opportunities as well as farm organization and practices. The designing of soil and water conservation policy for the 1980s and beyond must face this changed situation in contrast to the 1930s.

FARM MARKETS

In 1935 there were 6,814,000 farms in the United States. In 1980, there were 2,309,130. On the supply side of the market, the many sellers situation generally prevailed in the 1930s. On the demand side, the many-buyers situation was the general rule. Efforts to control supply were not widely practiced, nor were they successful. Market integration was not the rule, nor was dominance by a few large firms.

The market then was primarily domestic; market surpluses were the issue. Supply reduction by reducing production was held to be complimentary with soil and water conservation policy. Some payments to farmers to withdraw land from production had a primary objective of reducing supply and a secondary objective of conservation benefit.

Development of a conservation policy for the 1980s will be faced with a drastically differing situation and with trends probing in different directions. The maintenance of efficient production to sustain a large export market is deemed desirable from an agricultural income and balance-of-payments point of view. The resulting pressure on farm production has been to expand tillage to the limit on individual farms and to move onto land with higher erodability. Also, higher rates of fertilization and pesticide use and livestock concentration have increased concerns for surface and groundwater contamination.

Other agricultural products are not in this pattern. For example, milk production is now in surplus, with culling the poor-producing cows to reduce herd size and total production being advised. How will such policy advice, however, feed back to effect land management and soil and water conservation? The point is that each agricultural product will have to be analyzed along with its marketing situation to devise profitable conservation systems.

The increased specialization of production, integration of production and marketing, and greater involvement of production credit with its dependence on land value inflation as a security all affect the farm operation. These factors are quite different today than in the 1930s and will be forces affecting workable conservation policies in the 1980s and beyond.

POLITICS OF AGRICULTURE

The organization of governmental agencies to carry out the conservation policy of the 1930s came from a Congress with a politically powerful farm block. Furthermore, the policy was implemented by state legislatures dominated by farm interests and agriculturally oriented local units of government. But in 50 years, population has shifted. Urban interests control most state legislatures, and even many local governments that include a strong agricultural economy may not be primarily farming in point of view. Even the farming point of view has changed.

In the 1980s, agricultural interests will have to be astute coalition builders to achieve results. If these results depend on sizeable appropriations (for example, for cost-sharing of conservation practices), they must be won in strenuous competition with other claims. This does not mean, however, that a public interest in soil and water conservation does not exist. The 1970s saw the development of a widespread and strong environmental movement. The composition of this movement is not synonymous with the conservation movement, although there is major overlap (for example, shared concern for sediment in lakes and streams). Nonetheless, the development of soil and water conservation policy must take into account the existence of both interests. Achievement will depend upon creating areas of agreement and complementarity of purpose.

Two added points must be taken into account: First, urban erosion control and water conservation are major issues of recognized importance. Second, the population migration to country living means that more nonfarm people are living in rural areas. These facts do not mean that interests are necessarily congruent. Political insight and skill will be required to achieve a policy reformulation. The politics of the 1980's and 1990's need careful assessment at local, state, and national levels.

PUTTING CONSERVATION ON THE LAND

The policies of the 1930s were made in Washington, state legislatures, and the local district or committee with the recognition that the real action had to take place on the land with a change in farm practices. All of the administrative superstructure and rhetoric is of no avail unless erosion is reduced and water conserved on the land. Thus, the outcome of policies of the 1930s was to provide technical assistance to the farmer, cost-sharing for conservation practices, the authority to make land use rules and regulations in a few cases, educational programs to demonstrate which practices to follow, and the financial support for research under a wide variety of geographic conditions throughout the nation. All of these activities were to come to bear upon the individual decisionmaker and his/her use of land and water. The necessity to bring about change on the land has not shifted between the 1930s and 1980s.

These objectives of getting practices on the land have been fulfilled in part. But the plateauing effect previously mentioned is real and probably a factor in achieving funding goals agencies have thought desirable.

To achieve the objective of getting the owner/operator to put the practices on the land, to reduce erosion, and to conserve water will require a clear characterization of the owner/operator's whole decision unit and the forces motivating it. These characteristics are beginning to emerge from social science research initiated during the past decade. Yet we have only a rudimentary understanding of the relationship of owner/operator's personal and situational characteristics. How does the changing structure of agriculture and agrarian ideology alter soil conservation practice and policies? How do they affect the adoption of conservation practices? Many of the altered forces playing upon the owner/operator have been identified. But the character of the whole decision unit itself is continually changing as the business of farming changes.

The following points should be kept in mind in developing national policy and how it might influence the owner/operator of the 1980s:

1. The economics of agricultural production is dynamic, thus continually changing the character of the business.

2. The methods of production are likewise dynamic, with the prospects for the next 25 years being no less subject to change than the last 25 to 50 years.

3. The nation is very diverse in agricultural enterprise, land characteristics, climatic characteristics, etc. Conservation policy must be flexible enough to accommodate this diversity.

4. The decision unit makes decisions in a complex way, taking many factors into account that in turn, change over time. Thus, the influence of peer pressure in the 1930s and 1940s will be quite different in the 1980s and 1990s. Not only has the community of peers changed, but possibly the way in which the ethical considerations internal to the decision unit come to bear on the decision, For example, is the "force" of a conservation ethic different when expressed through a corporate decision unit, albeit a family corporation, than through a family with a patrician as head and a child or child-in-law in line of succession? How much action will take place as a result of technical assistance and education? At a time when regulation is increasingly viewed with disfavor, will regulations be necessary to achieve the less than 5 tons per acre per year of soil loss goal or some other criterion of performance? Is it a reasonable goal? Can erosion be shown to be detrimental to the income of a community, thus supporting local regulation and stronger peer pressure to the extent the peers are in the community? Must primary reliance be placed upon new conserving practices that are economic or new combinations of practices that will become feasible in the 1980s and 1990s?

The organizational structure to achieve the purposes defined in the 1930s still largely exists. New organizations with redefined goals are not created de novo, for the inertia to change must be taken into account. Is it possible for the present organization to become an agent for changing policy rather than a conservator of past modes? To what extent has it played such a role?

THE PUBLIC INTEREST

The issue of accepting conservation of land and water as a public responsibility, at least in part, was important in the 1930s as well as the 1980s. Arguments ensue as to how much of a public responsiblity. What is an appropriate level of federal investment for soil and water conservation? What criteria can be used to help assess which levels are most appropriate? What are the owner/operator's goals, and what do they consider important?

Also, critical questions have been raised as to whether government should cost-share a farm practice that is primarily profitable to the farmer. Do owner/operators strive for only economic viability and secondarily consider conservation? Would the owner/operator have followed the metric without cost sharing? What level of cost sharing is appropriate for which practice? Are the results of cost sharing tangible or intangible? Is the future benefit from reducing erosion today capitalized into farm values? Is the public responsiblity different for the family farm than for the corporate farm? Questions such as these have become more common, persistent, and need answers. Yet the value judgement has been broadly accepted that a portion of the benefits from erosion control and water conservation accrue to beneficiaries off site and at some future time. Therefore, some level of public expenditure is justified.

This basic perception remains, but an increasingly critical look is being taken to insure that public expenditures are securing conservation actions that would not have been taken otherwise and that actually result in less erosion, less pollution, and more conservation. In addition, there are those who currently argue that soil and water conservation is of such a high priority that benefits from other farm programs should be limited until compliance can be established with prescribed conservation practices. This concept has not yet passed the test of political and administrative acceptability, however.

POLICY RESEARCH NEEDS

We conclude that soil and water conservation policy for the 1980s and 1990s will be developed in a markedly different political, economic, social, and technical context than that of the 1930s and succeeding years. Because of the magnitude of these differences, research must be directed at more than incremental change and tightening up of existing programs. Fresh possibilities must be taken into consideration while maintaining an awareness of existing organizations and the issue of transitional change. Taking any less than a broad perspective will limit the opportunity for innovation and conservation improvement.

Research can explore policy alternatives and the forces that will be shaping their considerations. It can also improve understanding of the role of conservation in daily agricultural practice and give direction to biological and physical science research. In suggesting future lines of research, we are mindful of the previous discussion and we shall assess the proposed future lines of research in terms of our judgement about the context within which policy of the 1980's will be formulated and the anticipated direction of tomorrow's agriculture.

The proposed program of research has three key elements: (1) Institutional options and performance evaluation of alternative policy instruments, (2) the influence of policy instruments on farm adoption rates of soil conservation practices, and (3) policy and institutional studies growing out of the indirect effects of soil conservation on water resources.

A premise for indentifying the following research needs is that policy choices are being made, at every level of government, affecting the content and scope of conservation programs. There are choices between conservation and other public investments. Within conservation, there are choices among major instruments (education, tax incentives, technical assistance, regulation, cost sharing, cross-compliance), techniques, and institutional arrangements. These choices will be made. We prefer that they be made on the basis of valid information rather than historic precedent or "gut feeling." An essential research need is to provide information on the implications of relevant soil and water conservation policy alternatives.

The institutions and implementation strategies that took form in the 1930's are coming under increasing criticism. At the same time, there has been experimentation in a number of states with new programs and revised forms and powers of local conservation institutions. We propose to take advantage of these possibilities. The opportunity exists to evaluate a number of the promising institutional and policy departures in terms of actual experience in states and districts. information system is essential.

There is also an intermediate need between the making of administrative or legislative decisions and research results. Professional analysis for policy must assist the decision process. Such decisions often must be made in very short time periods. To facilitate these decisions, the compilations of data and research results must be analyzed into a short succinct decision document by highly competent policy analysts. This analysis depends upon the availability of coordinated, consistent data bases for the analyses of resource problems and the capability to consolidate and translate research findings into a form and context that is useful to decision makers. The policy analysis will be carried out by many people within the USDA, universities, and private research foundations. The results of this analysis will be consistent only if the data used are consistent. To aid in this task, a coordinated information information system is essential.

RESEARCH NEEDS

To help governments and citizens design and implement effective conservation programs based on knowledge of alternatives and evidence of their relative performance, the following question must be raised. What difference does it make whether conservation option A is chosen over option B? A sound evaluation of this question requires the development of research programs that analyze information from a range of subject areas. The programs involve three major components, along with a number of key elements discussed below. The research to tackle these needs should keep in focus the high risk areas for soil and water conservation. The basic criterion in delineating high risk areas is the annual impact of agricultural production activities on the future productive capacity of the land and water resources.

Research Need I: Institutional Options and Performance Evaluation of Alternative Policy Instruments.

Element 1.1 Political Setting for Conservation

Less than 3 percent of the national electorate is directly involved in agricultural operations. This fact must be recognized in assessing the political support for conservation programs. Such support reflects the public's perceptions of the role of soil and water conservation in sustaining the nation's well-being. Non-farm pressures are increasingly impinging upon agricultural lands and off-farm impacts of agricultural activities are becoming more sharply defined. As these changes progress, the political constituency supporting conservation programs may change along with the types of strategies advocated. Gaining a better understanding of these changes is an important research task.

The perception that food and fiber production are basic to our national interest is becoming more widely recognized. The long-term maintenance of the nation's resource base is vital to assure that both domestic and international food and fiber needs can be met at affordable prices. Seen in this light, agricultural lands are a strategic economic resource. Furthermore, the maintenance of water quality and quantity is also important if the nation is to realize the full range of potential benefits from the land.

A basic goal of research on the political support base for conservation is to gain a better appreciation of the relative importance that should be given to attainment of the many different social objectives influenced by conservation activities. These would include, but not be limited to, soil and water quality and quantity, consumer food costs and quality, minimizing the risks of future shortages of food and food price volatility, and farmers' production costs.

Element 1.2 Institutional Option

Two categories of institutional alternatives must be considered. The first involves a vertical slice through the agency structure—federal, state, and local. The second category considers various institutional options at a particular level of government. This horizontal comparison at the local and state levels is designed to examine the implications of various district organizations and state governmental arrangements.

A. Vertical structure. At issue in this set of vertically sliced institutional options is the allocation of discretion among levels of government. Discretion is exercised in setting conservation priorities and spending scarce funds. Examined options could include:

1. Block grants for soil and water conservation from the federal level to states to be allocated by the states. The formula for fund distribution among the states is clearly an important variable.

2. Block grants to the soil conservation districts from the state, with spending discretion at district level.

3. Matching fund approaches between state and district.

4. Local districts may contract for conservation practices installation, technical assistance, education, from any supplier.

B. Horizontal structure. The focus of this research is on the structure of local institutional arrangements and the performance differences attributable to structural differences. Alternative arrangements include:

1. District as a part of general local government, funded by direct state appropriation.

2. Independent local districts with local power to tax.

3. Differences in eligibility for district membership and election, for example, land ownership as a condition of membership.

4. Decisions on cost-sharing made by the district alone, by the Agricultural Stabilization and Conservation Service county committee alone, or by both.

5. Initiatives by the district as functions of various factors that describe the social setting for the district (population mix, economic structure, type of agriculture, land onwership pattern, etc.).

Element 1.3 Policy Instruments

The several instruments for influencing the conservation decisions of individual land managers must be compared. These include cost-sharing, regulation, technical assistance, tax incentives, cross-compliance, and education. Each implies an implementation cost and differing distribution between land managers and taxpayers. Performance impacts of each instrument may be identified generically and then further clarified by case studies. Certain adjustments in these instruments must be posed (for example, change in cost-sharing rate) to observe the sensitivity of performances to these changes.

Element 1.4 Performance

As with institutional options, policy instruments will be compared to determine any performance differences that may be linked to structure.

Performance indicators for both options and instruments extend beyond conservation practices implemented and include: A. Final products are:

1. Reduced soil movement.
2. Effect of reduced erosion on crop production, reduced cost of clean-up (ditches, streams), and off-site effects on water quantity and quality.
3. System-wide effects, for example, food prices.

B. Distribution of costs and results among relevant categories as:

1. Farmers, consumers, input manufacturers.
2. Income groups.
3. Commodity types.
4. Geographic region.

Research Need II. Farmer Adoption of Conservation Practice

Quantification of the performance measures for final products (reduced soil movement, crop yields, off-site water quality, food prices, etc.) and the distributional components (consumer welfare, farmer incomes, input manufacturer margins, etc.) and their relationship with alternative policy instruments requires an information support system on land use and conservation. This information system should focus on the behavior of those who control the land in high risk areas. This information should include the age, farm size, capital and equity position, intergenerational transfers, educational level, and managerial skills of landowners and operators.

The principal measurements that are emphasized for each dimension are the relationships between policy instruments and adoption rates of soil conservation practices and the indirect relationships of policy instruments on areawide soil and water quality. To capture these relationships and to evaluate the effects of soil conservation policies on the adoption rate of soil conservation practices and their influence on productivity, four research elements are identified.

Element 2.1 Effects of Soil Erosion on Agricultural Productivity

Agricultural productivity is mentioned by other panels. We repeat it here to give logical meaning to our proposed research and to place it in the policy context.

Public expenditures in conservation research and programs should wherever economically feasible be guided toward sustaining and enhancing the productive capacity of the natural resource base underlying agriculture. Research is needed to extend and define our ability to identify lands subject to declining productivity from soil erosion. It is equally important that this knowledge be predictive so that corrective actions in the form of conservation practices and systems can be installed on lands threatened by erosion. Two research activities are required to assist program administrators in identifying high risk areas. Using existing data sources and drawing upon our current understanding of the erosion-productivity interrelationship, a method to predict the impact of erosion on crop yields should be developed. Recent work completed as part of the RCA study is an example of the type of analytic model that would serve this purpose. The needed model would be nationwide in scope, would build upon the soil mapping unit as the basic unit of analysis, and would cover major soils-crops-management system combinations. Past research in a number of disciplines could be drawn on to prepare a preliminary list of high-risk areas in which more detailed work could begin.

Once available, a set of erosion-productivity relationships would rely on two research activities to improve model accuracy and reliability. First, inconsistancies in the model's data base should be corrected through field-level accuracy checks, especially for the estimated crop yield and soil depth information. Second, basic research should progress on the underlying physical relationships between erosion and productivity, with special emphasis on developing widely applicable empirical measures of complex soil-crop interactions that effect land productivity and can be integrated into the more aggregate erosion-productivity model.

Knowledge of the impacts of erosion on productivity would be useful for purposes other than identification of high risk areas. Quantification of the impact of erosion on crop yields from a given field each year would be valuable in calculating the on-farm social cost of erosion losses. This valuation of erosion would, in turn, be helpful in establishing public cost-share rates or other incentives offered as part of conservation programs. Furthermore, once high-risk areas are known, the efforts of scientists working on conservation technologies can be concentrated on soils subject to declining productivity under current technology and cropping patterns.

Element 2.2 Farmer Perceptions of Soil Erosion on Productivity

To design policies that will bring about greater use of conservation practices, it will be necessary to understand farmer motivations and behavior. Prior research suggests that farmers are adept at judging what is in their own current self-interest and at adopting courses of action consistent with that self-interest. If their perceptions of the relationship between soil loss and productivity is different from the relationship suggested by research, apparent irrational behavior may turn out to be quite rational indeed.

Knowledge of the difference, if any, between farmer perceptions and actual soil loss would aid in dealing with two issues. It would serve as an explanation for farmer behavior, and it would guide educational efforts at the farmer level–what education is needed, if any.

Element 2.3 Factors Affecting the Gap, If Any, Between Measured and Perceived Effects of Soil Erosion on Productivity

The gap between measured and perceived effects of soil loss on land productivity, if it exists, is influenced by a number of factors that can be classified into four groups.

A. Societal context. The perceptions of the community in which the farmer lives are important. His own perceptions will be either reinforced or opposed by various segments of his community. His neighbors, landlords leasing him farmland, and even farm equipment dealers are among those influencing the degree to which such a gap exists.

B. Individual characteristics. Research has shown that individual characteristics of farmers influence the adoption of farm practices. However, this research has only begun to address erosion control practices and has suggested that differences might exist in the way farmers adopt soil conservation practices as opposed to other kinds of innovations. These factors also would likely influence the gap between measured and perceived effects of soil loss on land productivity. Studies in this area should attempt to better understand the relationship of farmers characteristics (age, education, income, and farmer ideologies), size of operation, tenure, business organization, management style, and access to farm information to the degree of the gap between measured and perceived relationships.

C. Economic context. A number of economic influences could explain the existence of a gap between measured and perceived effects of soil loss or crop yields. For example, land market prices reflect not only its inherent productivity value but, in addition, its store of value as a hedge against inflation. In addition, credit markets have become increasingly unstable, and interest rates are at significantly higher levels.

Under the above influences, economic signals lead to a heavy discounting of the future by individual landowners and operators. As a result, the incentives for landowners and operators to adopt new conservation practices are weak. Moreover, under such incentives farmers are not likely to be motivated to collect information and data on the relationships between soil loss and productivity. As a result, economic signals may lead to the gap referred to under 2.3.

D. Conservation practices. Studies on the variations in soil loss by conservation practices is important in terms of its influence on farmers' perceptions of soil loss. For example, farmers' perceptions of conservation practices may be more in terms of economic benefits than effects on soil loss. Double cropping of wheat and soybeans in Missouri, while reducing erosion, may have been adopted more on the basis of economic incentives. Research in this region might address such questions as these: What is the relation between multiple cropping and adoption of minimum tillage?

Would incentive to stimulate double cropping increase the adoption of conservation tillage? What geographical constraints exist in adoption of multiple cropping? How do all of these influence the gap between farmers perceptions and measured soil loss on land productivity.

Research dealing with this element must, by nature of the problem, be site specific and multidisciplinary in nature. A research emphasis focusing on high erosion areas will provide a desirable setting for this kind of research. It will provide a specific context for the multidisciplinary effort to be undertaken. The STEEP (Solutions to Economic and Environmental Problems) project is such a multidisciplinary project in the Pacific Northwest focusing on soil erosion. It was funded in 1976 as a long-term project (15 years), primarily as a result of a strong farmer involvement. Research in the areas of tillage; breeding, diseases, insects, and weeds; runoff and erosion; and socioeconomics of soil erosion are included.

Element 2.4: The Effects of Fiscal, Monetary, Sector (Agriculture and Energy) and Soil Conservation Policies on the Adoption of Conservation Practice

General economic policies, fiscal and monetary, influence (among other things) the rate of inflation, interest rates (real and nominal), and the export demand for U.S. food and fiber. The current rate of inflation, along with interest rates, creates incentives for soil exploitation. Even in the face of decreasing soil quality, land prices continue to increase at a rate that exceeds the rate of inflation. Some agricultural policies, including price supports and deficiency payments, have equivalent qualitative effects on the price of varying qualities of land. Energy sector policies have a bearing on the production cost structure and thus influence the cash flow of farmers, which, in turn, affects the adoption rate of conservation practices. All of these policies influence farmers' time preferences, discount rates, relevant length of planning horizons, etc. and thus the rate at which they adopt conservation practices. Moreover, the influence of soil conservation policies (income tax incentives, direct subsidies, taxes and standards, cost sharing, low-cost loans, cross compliance with sector policies, extension and technical assistance, etc.) will be determined in large part by general economic (fiscal and monetary) and sector policies. In other words, the effectiveness of such specific policies on the rate of conservation adoption depends critically on more aggregate economic and sector policies.

To adequately respond to (2.4) we must:

A. Determine the qualitative and quantitative effects of general economic, sector, and conservation policies on rates of adoption.

B. Rank the relative importance and the interdependences among the various policies on rates of adoption.

C. Determine how soil conservation policies should be conditioned upon general economic and sector policies (treated as predetermined) to achieve alternative rates of adoption.

Research Need III. Policy and Institutional Studies Growing Out of Indirect Effects of Soil Conservation on Water Resources

Water policy research is too broad to be encompassed here. However, soil erosion affects water flows and water quality significantly. Soil conservation policies need to be studied from the perspective of these impacts. The major impacts are as follows:

A. Sediment and chemical pollution effects of soil erosion or conservation.

B. Water yield and flow effects of soil erosion or conservation practices.

C. Impact upon groundwater quality and quantity under alternative farming and conservation regimes.

Policy research requires specification of these effects over areas larger than the farm field. We urge refinement of the models, such as CREAMS, and the development of correspondingly comprehensive models covering watersheds and basins.

Element 3.1 Effects of Alternative Soil Conservation Practices on Different Water User

Depending on their different location in the basin, their different tolerances for uneven flows and irregularly timed flows, and their different needs for water quality as opposed to quantity, irrigation, urban, industrial, recreation, and energy production, users may be affected very differently by hydrological consequences of soil conservation programs.

Element 3.2 Erosion Control–Water Quality Policy Options

In a wide range of situations, control of soil erosion is a major, and necessary element in the control of nonpoint sources of water pollution. The Environmental Protection Agency and USDA are funding a number of research projects to test alternative control policies. These studies need to be pressed farther to determine:

A. What standards are most capable of implementation, for example, soil management practices, T-values, receiving water standards.

B. Alternative policy instruments(element 1.3 above needs to be studied from this perspective).

Element 3.3 Institutional Options.

Control of water flows and water policy can proceed upon bases of constitutional and statutory authority and public support beyond what is available for soil conservation per se. Possible questions are as follows:

A. Insofar as elements of land use regulation may be required for soil erosion control, may this gain feasibility as part of water pollution control?

B. Are nonfarm publics more effectively enlisted in the support of conservation programs if these are identified as water quality, flood control, or other water programs?

C. Are local flood control, drainage, lake management, or watershed districts effective managers of portions of the soil conservation task? Element 1.2B needs to be studied from this perspective.

D. What are the policy outcomes of alternative vertical linkages from national to local government institutions when soil conservation is considered in the water resources context? The chain of decisions and intergovermental links may pass through different congressional committees, federal departments and agencies, state agencies, and local government units. Different voluntary groups are given access by these alternatives. What are the results in terms of resources invested and farmer adoption?

SUMMARY

In summary, we have observed that the conditions affecting soil and water conservation have changed dramatically since the nation's policy base was laid in the 1930s. To aid in charting the course for policy in the 1980s and 1990s, the foregoing research is deemed exceedingly important if soil and water resource assets are to serve the nation when the population will move from a 1980 total of 226,505,000 to 243,513,000 in 1990; 260,378,000 in 2000; and 300,349,000 in 2030. That is an additional 74 million people in 50 years.

REFERENCES AND NOTES

1. Arnold Miller, Impact of Expanding Agricultural Production on Soil Erosion: Land Resources, Conservation Practices and Policy Choices, prepared for the Structure of Agriculture Project, U.S.D.A.

2. Carlson, John E., D.A. Dillman, W.R. Lassey and M. McLeod. 1977. The Farmer, Absentee Landowners and Erosion: Factors Influencing the Use of Control Practices. Research Technical Cometition Report. Idaho Water Resources Research Inst. Univ. of Idaho, Moscow, Idaho pp 6-7.
3. In 1930, 17.6 percent of crops harvested were for workstock and in 1960 it was only 1.5 percent and even less today. (Prepared by Natural Resources Economic Division, USDA, from Census and crop reporting data.)
4. The use of agricultural chemicals increased from an index of 39 in 1930 to 129 in 1979. (Prepared by Natural Resource Economic Division, USDA)
5. Bureau of the Census, U.S. Department of the Interior
6. Bureau of the Census, U.S. Department of the Commerce, Current Population Reports,. Series P-25, Number 704, July, 1977. Medium rate projection needs.

6

Walter F. Megahan, Chairman

Don H. Boelter, Stanley P. Gessel,
James W. Hornbeck, James R. Meiman,
Gary R. Nordstrom, Edwin H. White

FOREST LAND

Conservation Needs, Technology, and Policy Alternatives

Forests exert a profound influence on watersheds. They grow, mature, and ultimately die following a successional pattern of species that change the microclimate, the soil, the water outflow, and even the topography of the landscape. In so doing, forests exert influences both onsite and downstream. Changes in forest stands, either natural or human related, can cause major changes in soil and water values as measured by: (1) onsite erosion and nutrient cycling; (2) the quantity, timing, and quality of streamflows; and (3) the riparian ecosystem. Inasmuch as 75 percent of the nation's water originates on forest lands, reductions in soil and water values on forest lands are extremely important.

Many forest activities included in producing and harvesting forest crops and forest uses such as mining, waste disposal, grazing, and recreation may adversely affect forest soil and water resources. In addition to man's activities, natural events such as wildfire, extreme climatic events, insects and diseases, and geologic phenomena affect quality of soil and water resources on forest lands. Maintenance or enhancement of onsite soil productivity and the quality and quantity of streamflows are primary concerns. Forest management decisions must provide a balance between the production of goods and services and the maintenance of soil and water values.

Forested lands have a substantially different set of requirements for soil and water conservation research than do croplands. Although concerns about site productivity are similar to those for agricultural lands, there is even greater interest in the offsite or downstream impacts. Flooding, sedimentation, low flows, pesticide transport, and damages to aquatic habitats are examples of these off-site concerns. Furthermore a significant amount of the soil loss from forested lands comes from a relatively small percentage of the area, and much of this soil loss is by gullying, mass failure (landslides), and stream channel erosion in contrast to the sheet and rill erosion on cultivated lands. The variety of land uses, and broad range of biologic, topographic, and climatic conditions make forest land management more difficult than for most agricultural areas.

CURRENT SITUATION

Present Conditions and Future Needs

A total of 737 million acres of land or almost 31 percent of the total land in the United States is forested. Of this, 65 percent or 482 million acres is classified as commercial forest land.[1] Forest lands are widely distributed and, with the exception of the Great Plains and some of the Southwest, compose a significant part of the area of each State. However, nearly three-quarters of the commercial timberland area is in the humid eastern half of the country and is about equally divided between the North and South. Commercial timber stands in the West are concentrated in the Pacific Coast States of Oregon, Washington, and California; and in the Rocky Mountain States of Montana, Idaho, and Colorado. The remaining 255 million acres of forest land is either incapable of producing a sustained crop of industrial wood or is reserved for special uses such as watershed protection, parks, or wilderness areas. (USDA Forest Service, 1980).

Approximately 451 million acres, or 61 percent of the nation's forest land, is under non-Federal ownerships. Forty-five percent of these forests are in the Southeast with most of the remainder in the Northeast. Most of these lands are small private ownerships representing a diversity of forest management objectives, widely different characteristics and attitudes, a limited knowledge of management opportunities, and varying willingness and capacity to make investments to preserve soil and water values.

Soil and water resources are already impaired on many forest lands. The prognosis for the future adds to this concern. Projected increases in population, economic activity, and income suggest ever-increasing demands on forest resources. For example, demands for timber are expected to increase by 64 percent by the year 2000 and 107 percent by 2030 as compared to 1976 levels (USDA Forest Service, 1980). Similar increasing demands are expected for all uses and products on forest lands throughout this period.

Accompanying the increased demand for the resources from forest land is a probable decrease in the area of forest land in response to a variety of demands such as agricultural needs, increased urbanization, transportation corridors, withdrawals for wilderness and water development. To illustrate, from 1967 to 1975, more than 100,000 acres of forest land was lost annually to alternative uses. Losses of forest land will accelerate in the future, especially on the 42 million acres of forest land classified as prime farmland (USDA Soil Conservation Service, 1978), as demands for agricultural and industrial products, including alcohol, increase.

Pertinent Legislation

The Organic Act of 1897 defined water as one of the primary reasons for establishment of the National Forests as follows:

"No national forest shall be established, except to improve and protect the forest within the boundaries, or for the purpose of securing favorable conditions of water flows, and to furnish a continuous supply of timber for the use and necessities of citizens of the United States."[2]

This concern was reinforced by the 1911 Weeks Law that stated: "The Secretary of Agriculture is authorized and directed to examine, locate, and recommend for purchase such forest, cutover or denuded lands within the watersheds of navigable streams as in his judgment may be necessary to the regulation of the flow of navigable streams or for the production of timber."[3]

Much of the early concern was based on the assumption that forests are necessary "...to prevent destructive floods and corresponding periods of low water" (Pinchot, 1947). Although subsequent research has refined our understanding of these relationships, the fact remains that early legislators recognized as a key issue the ties between forest cultural operations and streamflow.

A variety of subsequent legislation has been enacted to maintain and enhance the quality of soil and water resources on forestlands. Important acts for managing National Forest lands such as the Multiple Use-Sustained Yield Act[4], the Forest and Rangeland Renewable Resources Planning Act[5], and the National Forest Management Act[6] all include protection of soil and water resources as a central theme. Additional legislation including the Watershed Protection and Flood Prevention Act[7], The Resource Conservation Act[8], the National Environmental Policy Act[9], and the Federal Water Pollution Act [10] help to protect soil and water resources on all forest lands regardless of ownership. In addition, many states have adopted state forest practices acts that are directed at conserving soil and water resources.

Unique Character of Forest Lands

Forest lands are unique in many respects and thus pose special problems for management of soil and water resources. Most existing forests are on land not well suited for agriculture. The most common reason for this is topographical—the land is too steep or too dissected. This is almost universally true in the West and is widespread in the East as well. Steep slopes greatly increase erosion hazards and expand the spectrum of erosion processes. This topography also causes wide variations in microclimate.

Forest lands are usually utilized for long-term crops with rotation ages ranging from 25 to 100 years or more. Because of this, they are particularly vulnerable to long-term climatic trends. The longevity of forest stands increases the probability of damage from natural disasters — such as wildfire, insect and disease attack, and extreme climatic events.

Economic constraints influence the management of forest lands for soil and water resources. Investments in improved practices have been less attractive because of long harvest intervals necessary to recover funds. The opportunity for application of conservation practices has been limited, as compared with agricultural lands, but this is changing, and enhancement of productivity by fertilizing has been shown to be economic. Other values such as recreation may equal or exceed timber importance on forest lands, but they provide little income and investment opportunities to forest land owners.

Finally, forests cover a broad range of geologic and soil characteristics. These factors, coupled with extreme ranges in climatic conditions, cause wide variations in the biological community of forests. This extreme variation makes it difficult to generalize the results of observations and experiments dealing with the effects of forest practices on soil and water values without a careful evaluation of the conditions under which the studies apply.

STATEMENT OF PROBLEMS

Erosion and Sedimentation on Forest Lands

Magnitude of the Problem

Unlike agricultural lands where wind and surface erosion processes dominate, the most prevalent erosion processes on forest lands include surface erosion, a number of mass erosion processes and channel erosion. Transport of eroded material downstream as stream sediments is also unique on forest lands because of the influence of forest debris in channels.

After comparing all possible sources of pollution from forest lands, the EPA (U.S. Environmental Protection Agency, 1973) concluded that:

"Sediment, including both its organic and inorganic (mineral soil) consitituents, is the greatest single cause of water quality degradation. Control of erosion due to runoff is thus the most important aspect of control of pollution from forests."

Forested areas produce a substantial amount of the total sediment production for the nation. Fowler and Heady (1981) report that undisturbed forest lands alone contribute nearly 110 million tons of suspended sediment to the nation's public waterways each year...disturbances on forest lands can greatly accelerate these natural sediment yields. The northwest and southeast regions of the United States have the greatest suspended sediment production rate in the country. It is well to bear in mind that these data are for sediment production...not erosion. A large proportion of eroded material does not appear at downstream locations as sediment because of deposition enroute. Thus, suspended sediment data represent only a fraction of the actual watershed erosion. This fraction commonly ranges from 20 to 60 percent on the relatively small upland basins common in forested areas (Renfro, 1975).

The impacts from increased erosion on forest lands can be severe and long lasting because of downstream sedimentation effects. To illustrate, valuable chinook salmon spawning beds in the South Fork of the Salmon River in Idaho were literally buried under bedload sediment after logging and associated road construction caused excessive sediment inflows. Because of the damage, the entire upper watershed (260,000 acres) containing a total timber growing stock of over 3 billion board feet was closed to all logging and road construction. Studies of channel conditions showed that it took 12 years for the river to return to the point that the moratorium could be lifted (Megahan et al., 1980). The impacts of forestry on fisheries in general is now emerging as a very important problem, especially in the Pacific Northwest.

Thousands of water development projects are located in forested or predominately forested watersheds. Accelerated sedimentation caused by poor forest management or wildfire can lead to reductions in irrigation capacity, power production, or flood control benefits as reservoir storage capacities are lost to sediment. Associated losses of channel capacity in water conveyance works and in natural channels increases flood damages and often necessitates expensive remedial measures.

Erosion from forest lands and resulting sedimentation are important in other respects. Most forest soils contain a greater proportion of their available nutrients in the surface layers as compared to agricultural soils. Loss of a portion of elements essential to forest growth through erosion could affect future productivity. Many studies are in progress to evaluate nutrients lost by the removal of large volumes of forest products by intensive harvest practices. McColl and Grigal (1979) warn that more nutrients can be lost by erosion than are lost by the removal of large volumes of plant material. Erosion is important in another respect because other chemicals introduced into forests, such as pesticides, may become attached to soil particles which, appearing as sediments, could carry them into the aquatic ecosystem.

Processes

The spectrum of erosion processes is broader on forest lands than on agricultural lands. Surface erosion, including sheet and rill erosion processes, are common to both types of land. However, gully erosion, and especially mass erosion (landslides) and channel erosion problems are also common on forest lands. Delivery of eroded material to downstream locations is also different because of differences in vegetation and topography on forest lands and the effects of forest debris and channels.

Accelerated surface erosion following disturbance is a direct result of the nature and type of disturbance, not simply the removal of vegetation. Surface logging, as opposed to aerial logging activities, can cause accelerated erosion through disrupted soil structure, removal of protective cover, increased raindrop impact, and reduced soil infiltration rates that lead to overland flow of water. Forest fires, especially wild fires, can cause many similar problems, often to a greater degree. The changes in hydrologic function that lead to accelerated erosion are particularly severe on poorly designed roads. Roads have additional effects that tend to accelerate erosion further including (1) increased slope gradients on cut-and-fill slopes, (2) interception of subsurface water flow zones, and (3) concentration of overland flow of water on the road prism and in channels. Accordingly, roads often tend to cause a large increase in surface erosion per unit of land disturbed. Research has documented the fact that timber harvest activities, wildfire, and road construction can accelerate erosion to levels exceeding 10 or even 100 times those of undisturbed forest lands.

In addition to surface erosion, mass erosion (primarily in the form of landslides) can also accelerate in response to forest land disturbances. Vegetation removal, either by timber harvest or wildfire, tends to increase landslide hazards because of (1) increases in the depth of the zone of saturated soil in response to reduced evapotranspiration losses, (2) decreased vegetation root strength caused by root decay, and (3) increases in total snow accumulation and snowmelt rates. By increasing slope gradients on cut and fill slopes, removing support at the base of unstable slopes, changing soil strength parameters, and causing possible increases in groundwater levels, roads further increase landslide hazards compared to vegetation removal alone. Swanston and Swanson (1976) compared the incidence of 396 landslides on uncut forested areas, clearcut areas, and on forest roads on four different sites in Oregon, Washington, and British Columbia. The increase in landslide activity due to clearcutting alone ranged from zero to four times that for uncut stands for the four sites. Road-related failures amounted to 25 to 340 times greater than on uncut forest stands.

Fire may accelerate mass erosion of soil similar to increased erosion caused by other means of vegetation removal. Effects may be accentuated, especially in the case of intense wildfire, because all vegetation is removed without regard to slope gradient. Some of the most spectacular erosion events documented for forest lands result by this means. Post wildfire mass erosion is a major threat to property and life in southern California.

Sedimentation responses to accelerated surface and mass erosion vary depending on the type and severity of the erosion, the proximity of eroded areas to the watershed channel system, the nature of the climatic event causing the streamflow, and the hydrologic properties of the watershed.

Research Needs

There is a definite need to develop techniques to predict the erosional consequences of alternative forest management practices and wildfire. Such techniques are already available for agricultural lands. The Universal Soil Loss Equation (USLE) developed by Wischmeier and Smith (1965) provides an estimate of average annual sheet and rill erosion on agricultural lands resulting from rainfall. Subsequent work by Wischmeier (1975) and others attempted to adopt the USLE to woodland situations. Unfortunately, the data base needed to verify the role of causal factors does not exist for forest lands. Also, the USLE is not suited to many of the erosion processes occurring on forest lands.

Emphasis should be placed on more basic study of all erosion processes on forest lands that will lead to practical erosion control practices and physically based erosion models. This is particularly germane at the present time because Public Law 92-500 calls for controlling nonpoint source pollution on forest lands. The Environmental Protection Agency and the states are approaching this problem by developing guidelines for "Best Management Practices" for forest lands in order to minimize erosion to meet still-to-be-established water quality standards. A detailed knowledge of erosion processes and methods to control erosion is needed to establish "Best Management Practices" on forest lands — a level of understanding that does not presently exist.

Even more difficult is the problem of predicting the delivery of eroded material to streams and the subsequent downstream transport of sediment. This requires an understanding of sediment routing from the source to the watershed mouth. This phase of the sedimentation process is one of the least understood areas in forest hydrology and is vitally important because the time and space relationships between erosion and downstream sedimentation cannot be developed without it.

One other aspect of sedimentation from forest lands needs additional research. Most sedimentation studies to date have concentrated on the amount and size of mineral sediments. Yet, both mineral and organic sediments can serve as carriers for nutrients and pesticides. In addition, organic sediments constitute a large proportion of the sediment yield from forested areas and provide an important component in the aquatic food chain. Studies are needed to define the role of both mineral and organic sediments in the transport of nutrients and pesticides and the impacts of changes in these sediments on the aquatic ecosystem.

Quantity and Timing of Streamflow

Magnitude of the Problem

Forests and rangelands are the source for an estimated 70 to 80 percent of the nation's water supplies. In the 11 western states, over 90 percent of the usable water originates in high-altitude watersheds which are largely forested. Eastern and western municipalities utilize 2 million acres of forested watershed land as their source of supply or as protection for their reservoirs and well fields, including groundwater recharge areas.

At present there are water shortages at many locations, but especially in the "northeast corridor" and in the western United States. By the year 2020, the annual demand for water produced in forested regions in the United States is expected to increase by 30 million acre-feet in the East and by 39 million acre-feet in the West. Consideration and comparison of costs suggest that forest lands would be a productive and economical source of increased flows. Hibbert (1979) estimates that water yield in the Colorado River Basin can be increased by 750,000 acre-feet by a combined program of snow management, timber harvest, and chapparal conversion.

The same forested lands that serve as a water supply are also source areas for flood waters. The relationships between forests and floods have been argued for many years, but the current consensus seems to be that the forest is the best of all possible natural covers for minimizing overland flow, runoff, and erosion. However, there is potential for flood reduction from forests through continued fire protection, careful logging, and reforestation of abandoned lands.

Processes

Changes in forest stands, either natural or man caused, can cause major changes in hydrologic performance including the quantity and timing of streamflow. Over the past half century, process and watershed scale studies have been used to document these changes. Many studies have documented that cutting of trees increases annual water yields. Transpiration and interception losses to the atmosphere are reduced and are not matched by increased evaporation. As a result, soils remain wetter and more water is available for streamflow and ground water recharge. The increases in streamflow after cutting are generally proportional to the amount of cover removed. The relationship is linear and ranges from no increase in water yield if less than 10 percent of the forest cover is removed to increases of 10 to 16 inches in annual streamflow from forests that are clearcut (Douglass and Swank, 1972). In snowpack areas, a further increase in water yields may result from increased snow accumulation in openings caused by forest cutting.

Water yield increases usually decline rapidly upon regeneration of the new stand, and increases of more than one inch seldom persist beyond the 5th year of regrowth, except where streamflow is generated mostly by snowmelt. In the latter case water yield increases may persist for at least 2 decades (Leaf, 1975). Where hardwood forested watersheds are converted to conifers, water yield may eventually fall well below previous levels because of greater interception and transpiration losses (Swank and Douglass, 1974).

In most areas, forest cutting can be expected to increase streamflow in summer or autumn when augmented water supply is most needed. This increase is the result of reduced transpiration. An exception is areas where snowmelt contributes a large part of total runoff. In such areas, forest cutting may increase streamflow at the start of the melt season (Leaf, 1975).

Cutting or other disturbances of the forest have two important effects on storm and flood flows. First, soils in the cut or disturbed areas are wetter during the first few years after cutting, so there is less opportunity for precipitation storage. Second, disturbances to the forest floor and soil can change infiltration rates and may cause overland flow. Generally the changes in stormflow are not of major consequence to downstream flooding, particularly since the cut or disturbed areas are usually small in relation to total headwater area. At a local level the changes can be more important and must be considered in designing logging roads and drainage systems and in protecting the stream channel against erosion and sedimentation.

Removal or reduction in forest cover eliminates shading by stems and canopy, significantly advancing snowmelt and causing a desynchronization of snowmelt runoff between disturbed and desynchronized melt can change peak flows on smaller watersheds, but these effects are usually of negligible regional consequence. Only small patches of forest are cutover or disturbed at one time on major watersheds where diversity of aspect, slope, elevation, and vegetation already provide natural desynchronization.

Research Needs

Increased water yield resulting from forest cutting or other disturbances has been achieved and demonstrated at many places. Where and by what techniques can forest management produce the greatest or most economical yield of water? Maximum increases in water yield will result from elimination of forest vegetation that transpires at or near maximum rates. Favorable locations are areas that have deep soils, riparian zones where vegetation has access to groundwater, areas that have direct drainage to channels, and areas where streamflow per unit of precipitation is least (Anderson et al., 1976). These areas need to be indentified and suitable treatments devised for increasing water yield. Riparian zones need special attention. They are a fertile possibility for increasing water yields, but they are extremely fragile, and knowledge is limited. In conducting water yield studies, the impacts of management practices on aesthetics, wildlife, recreation potential, and water quality must be evaluated.

Foresters and others have gradually come to realize that forest lands are the best natural cover for protecting against rainfall floods especially in smaller basins. Research must focus on maintaining and even enhancing the flood-preventing capabilities of forests, especially during cutting or other disturbances. Control of runoff from disturbed surfaces such as roads, trails, and landings is a continuing research need.

Floods produced by spring snowmelt and rain on snow continue to be a major concern. Seemingly little can be done in forest management to prevent such floods, especially on a regional level. However, it is possible to influence flood flows in tributary streams and to better predict the magnitude of snowmelt-associated floods in all basins. Research is needed to develop procedures for accurately predicting the contribution of snowpacks to floods based on size and condition of pack, prevailing weather conditions, physiography, and forest vegetation patterns.

A number of simulation models exist that are applicable to specific regions. These models enable managers to estimate land use impacts in the hydrologic regime as well as to identify the weak links in our understanding of hydrologic processes. Specific study needs dealing with the effects of forest management on water yield and flooding are:

1) effects of various methods and patterns of vegetation removal on evapotranspiration and snow accumulation and melt;

2) the role and function of variable source areas;

3) disturbance effects on runoff processes including overland flow, subsurface flow, and flow routing through the soil mantle.

Water Chemistry

Magnitude of the Problem

In addition to sediment, Public Law 92-500 covers the chemical constituents of streamflow from forest lands. A variety of chemicals from natural and man-induced sources are cycled through the forest ecosystem and influence the quality of water runoff from the forest. Fores. practices can change the nature and rate of the cycling process and thereby affect water quality. Future pressures on forest resources will inevitably lead to a need for more intensive harvesting practices accompanied by an increased concern for water chemistry.

Streamflow from most forested areas is relatively low in natural chemical constituents and contains little introduced chemicals. The aquatic ecosystem has developed accordingly and is thus sensitive to changes in water chemistry.

All types of pesticides including herbicides, insecticides, fungicides, rodenticides, piscicides, and bird repellents, are being used on forest lands in the United States to some degree. Herbicides and insecticides probably account for close to 90 percent of the use. Fertilization is becoming increasingly important as a means to increase forest productivity in the United States. Through mid-1978, about 1.2 million acres had been fertilized in the Pacific Northwest. At this level, the West leads the country in fertilizer use. The only other region of heavy use is the Southeast where a total of 0.8 million acres had been fertilized by mid-1978 (Bengston, 1979). Of all the possible kinds of pollutants, fire retardants are unique to forest lands. Probably over 85 percent of the fire retardant use in this country takes place in the western states. In the bad fire year of 1977, the Forest Service alone used 14.5 million gallons of fire retardants.

In the past decade, acid rain, the popularized term for atmospheric deposition, has become a significant concern. Acid rain is a complex problem that is hard to define in terms of sources and impacts. The gradual impacts of acid rain are extremely difficult to quantify. Major concerns are an increased rate of acidification of soils and lakes and possible damage to the cuticular layer protecting foliar systems. The acid rain problem, which is currently of greatest concern in the eastern United States, is receiving more research effort.

Processes

Natural Chemicals. Forest cover helps regulate stream chemistry by storing nutrients in the standing vegetation and forest floor, and by influencing the flow of water through the soil profile. Soil disturbances and cutting of the forest can interrupt the nutrient-regulating ability and bring about increased leaching of ions into streams. An extreme example was a forest clearing study at Hubbard Brook Experimental Forest in New Hampshire (Likens et al., 1970) where nitrate concentration in stream water rose 50 times on the average and reached a maximum concentration of 80 mg/liter, and major cations increased 3- to 20-fold.

The Hubbard Brook study stimulated research on nutrient outputs in streams from commercially cut forests at many locations in the United States. These studies indicate that forest cutting, and especially clearcutting, will usually change stream chemistry, although the amount of the change can be highly variable. Stone (1973) suggested that the consistently greater changes obtained following commercial clearcutting in New England versus other locations in the country may be caused by such factors as: natural soil fertility; mineral soil depth; depth and type of organic accumulations; soil acidity; precipitation; and species, timing, and distribution of plant regrowth. Whatever the causal factors, the studies point to a continuing need to quantify changes in stream chemistry according to both geographical location and environmental parameters.

Introduced Chemicals. Chemicals including pesticides, fertilizers, and fire retardants are introduced into forests to reduce mortality of desired tree species, to improve forest production, to favor particular plant species, to control wildfires, and to increase water yield. In the future, more intensive forestry will undoubtedly require greater use of such chemicals.

Detection and evaluation of treatment effects is somewhat easy in the West because many forested areas are located in stream head-waters. Thus, water is essentially free of synthetic pollutants in the absence of chemical applications. By contrast, in the East many forests are intermingled with farms and municipalities so chemical pollutants are often introduced from other sources.

There are five major pathways for introduced chemicals to enter streams: direct application, drift, mobilization in ephemeral channels, overland flow, and leaching (Norris, 1978). Direct application is often the most important source of increased chemical concentrations. Concentrations tend to be highest from this source, but the duration of entry is short. For a given rate of deposition, the concentration varies with amount of water surface exposed and the volume of water. Chemicals can drift into streams from aerial applications on nearby areas. The amount of chemical supplied by this means is limited by the amount of drift; so concentrations are lower compared to direct application to streams even though the duration of entry is the same.

The remaining pathways are closely related to the hydrologic processes of the watershed. The actual amounts and timing of chemical inflows depend on the fate of the chemical enroute and the mode of transport. Oftentimes, ephemeral channels are inadvertantly treated during aerial applications. Flow may occur in these channels with a large enough climatic event, thereby mobilizing any residual chemicals. Similar effects may be observed if overland flow of water occurs outside of ephemeral channels, although this is rare on forest lands. Generally, water percolates through the soil on forest slopes and moves slowly downslope as subsurface or groundwater flow. Chemicals may be leached into streams by this means, but most chemicals don't persist long enough for this to occur. Any chemical increases occurring by this means would be low level and distributed over longer time periods.

Research Needs

Natural Chemicals. There are at least three major questions about the impacts of increased ion concentrations in streams. First, do the increases approach or exceed standards established for beneficial uses of water and protection of aquatic life? Second, how important are increased leaching losses in terms of site productivity? And third, what are adverse and beneficial impacts on stream biota?

According to currently available information, the nutrient concentration standards established for protecting water use and aquatic life are not being exceeded after heavy cutting. However, there are certain situations and locations where there is potential for exceeding standards. For example, a study on undisturbed watersheds in the southern pinery has shown that mean annual concentrations for total phosphorus in solution approach the 0.03 mg/liter benchmark level recommended for the protection of aquatic life (Duffy et al., 1978). Harvesting in these forests could cause phosphorus concentration to exceed recommended levels. A research need is to identify these situations and locations so that precautions can be taken.

Relationships of changing nutrient concentrations to site productivity and stream biota are only now beginning to receive needed research. The site productivity problem goes well beyond the solution losses in streamflow and includes such items as nutrient removal in harvested products and nutrients lost in sediment. The relation of nutrient removal and productivity is considered in other sections of this paper.

Introduced Chemicals. Research will be needed to determine the most efficient types, amounts, and methods of chemical application that will not exceed water quality standards or harm forest riparian habitats. It is important to bear in mind that all of the studies of the effects of pesticides on water quality were directed at the concentration of pesticides in water; pesticides attached to sediments were not studied, nor were the effects of the pesticides on forest riparian habitat conditions studied. Further, it is interesting that the three highest recorded outflows of natural chemical constituents were obtained when herbicides were used either alone or in conjunction with forest cutting. Studies are needed to better define the interactions between herbicides, nutrient outflows and on-site productivity.

Acid rain poses a potential threat to forest productivity, water quality, and riparian ecosystems. Continuing industrialization and increasing emphasis on coal for fuel may increase the acid rain problem. Funding has been made available and research programs are being developed, but many scientists express uneasiness about finding useful answers to questions on acid rain. Traditional field and laboratory studies and computer simulations are giving glimpses of causes and potential impacts, but radically new approaches and concepts will be necessary to unravel the perplexing questions.

Not all changes in water chemistry are necessarily bad. In some instances, productivity of the aquatic ecosystem is low because of limited amounts of dissolved nutrients. Studies are needed to better define the nature, extent, and duration of changes in water chemistry on the components of the aquatic ecosystem in order to better define limits of acceptability of changes.

The Riparian Zone and Acquatic Ecosystem

Magnitude of the Problem

The riparian zone is a streamside ecosystem that has a high water table because of proximity to surface or subsurface water. Riparian ecosystems are most commonly recognized by bottomland, floodplain and streambank vegetation. They are uniquely characterized by the combination of high species diversity, high species densities, and high productivity. Continuous interactions occur between riparian, aquatic, and adjacent terrestrial ecosystems through exchanges of energy, nutrients, and species.

The riparian zone represents a small fraction of the forest area in the U.S. and yet is recognized as one of the most important and valuable landtypes because it: (1) plays a critical role in determining the quality of the aquatic habitat for fish resources; (2) provides a vegetative buffer zone to help filter pollutants and provide stability during flood flows; (3) provides important habitat for terrestrial wildlife populations; (4) has a high aesthetic value; (5) is a focal point for many recreational activities, and (6) is a highly productive zone for both timber and range resources. Effects of disturbance of riparian zones have most often been evaluated in terms of impacts on wildlife, including birds, mammals, and fish. Fish are particularly vulnerable because they are endemic to the area affected. In spite of extremely high values, over 70 percent of the original riparian habitats in the U.S have been lost. Much of the loss has been through nonforestry practices such as channelization, farming, grazing, and flood control. However, forestry practices have also contributed to the problem through timber harvest activities, fire, and road construction. Impacts are caused by direct encroachment in the riparian zone and by changes in streamflow quantity and quality caused by operations elsewhere in the watershed.

Salmonid fish, including trout, salmon, and char, spend all or part of their life cycles in headwater forest areas and thus are sensitive to changes in aquatic habitat conditions caused by forest management practices. Successful survival and health of these fish are heavily dependent on the maintenance of good aquatic habitat conditions in headwater streams. Fish values alone can be sizable. For example, the commercial salmon catch for the west coast amounted to over 250 million dollars in 1974, and the sport fishing catch in Alaska alone was worth over 50 million dollars in 1973 (Meehan et al., 1977).

Processes

The U.S. Forest Service, in conjunction with university specialists, summarized the research needed to protect aquatic habitat conditions in the western United States (Meehan, 1975). Important habitat properties that can be affected by forest management practices were identified in the report, including streamflow and sediment rates, levels of natural and introduced chemicals, oxygen content of water, water temperature, and stream morphology. All of these properties can be influenced by direct encroachment in the riparian zone. Streamflow rates, sedimentation and water chemistry can also be affected by forest practices elsewhere on the watershed. Direct encroachment occurs because the riparian zone is often one of the most productive timber sites available. Consequently logging, including clearcutting, has often occurred in the riparian zone. Direct encroachment also occurs when the riparian zone is used for forest road construction, a common practice in mountainous areas because of more favorable grades and ease in construction.

Vegetation, predominantly woody vegetation, is the primary factor contributing to the unique character of the forest riparian zone. Standing woody vegetation provides cover to help regulate temperatures in these headwater streams and also serves as a buffer zone for inflow of pollutants from adjacent watershed lands. Intricate root systems stabilize banks and provide cover for fish. The vegetation is a source for insects and, most important, woody debris. Once in the channel, woody debris serves as the base of the food chain and provides habitat for lower life forms and fish. Debris also greatly influences channel morphology and bank erosion and is the primary factor regulating the amount and rate of bedload sediment movement out of forested watersheds.

Research Needs

Certain water quality parameters are well defined with respect to impacts on the aquatic ecosystems. For example, the physiological tolerances of fish for temperature, dissolved oxygen and pH are well defined. However, the effects of sediment, the most serious and widespread pollutant on forest lands, are relatively unknown. This is in spite of the fact that sediment can seriously impact every component of the fish life cycle including reproduction, rearing, and migration.

It is important to keep in mind that the real goal of the Clean Water Act (Public Law 92-500) was to create fishable, swimmable water. Each state has developed water quality standards to direct the country toward reaching this national goal. Unfortunately, the current water quality standards for sediment, for the most part, cannot be related to the general fishable, swimmable goal. The standards are based on arbitrary sediment concentrations or, worse yet, turbidity levels. Definitive studies of the effects of sediment on fish and other aquatic biota are needed to provide a sound technical base to establish workable sediment standards to meet national goals. We need to know how increments of sediment affect fish, and the impacts of timing, duration, and character of the sediment on the aquatic life systems found on forest lands.

Studies are also needed to delineate the effects of other water quality parameters on the aquatic ecosystem including changes in dissolved nutrients, temperature, and introduced chemicals. Studies should be designed to detect both the positive and negative aspects of changes in pollutant loading.

Direct encroachment on forest riparian vegetation by logging, wildfire, and road construction causes profound changes in the normal function of the riparian zone for filtering pollutants, regulating stream temperatures, providing insects and cover for fish, and stabilizing streambanks. Also, the cycling of woody debris to channels is affected. Yet, almost nothing is known of the consequences of these changes on aquatic biota and the routing of sediment. Studies are needed to define the nature, magnitude, and duration of these effects in response to alternative types of disturbances in the riparian zone.

Maintenance and Enhancement of Forest Productivity

Introduction

Resources Planning Act assessment of forest land shows that potential productivity varies widely.[11] Soil, moisture, elevation, and topography not only influence the kind of vegetative cover but also the amount of wood fiber or herbage produced at a given location.

Actual productivity is generally less than potential because of many ownership objectives and current stand conditions. Future productivity under intensive management systems could be significantly increased. The potential wood fiber production on all forest land in the nation averages 53 cubic feet per acre per year (USDA, 1980a). But the potential varies significantly among regions as well as within regions. In the East, the average potential is 68 cubic feet per acre per year. In the West, it is 37 cubic feet per acre per year, ranging from 5 cubic feet per acre per year for the pinyon-juniper cover type to more than 200 cubic feet per acre per year for the Douglas-fir cover type (USDA, 1978).

In the Northeast, about 8 percent of the forest land is capable of producing more than 120 cubic feet per acre per year, nearly 27 percent can produce 85 to 120 cubic feet, and 36 percent can produce 50 to 85 cubic feet per acre per year. In the Pacific Coast region, 31 percent of the forest land is capable of producing more than 120 cubic feet per acre per year. In the South, only 6 percent of the forest land is capable of producing at this level.

The average potential productivity of commercial forest land under forest industry ownership is 87 cubic feet per acre per year. For this ownership group, the average potential production in cubic feet per acre per year is 118 in the Pacific Coast region, 82 in the South, 75 in the North, and 74 in the Rocky Mountains and Great Plains. The average potential productivity for forest land under farm and other private ownership is 72 cubic feet per acre per year, with 100 cubic feet per acre per year in the Pacific Coast region, 75 in the South, 67 in the North, and 50 in the Rocky Mountains and Great Plains.

Magnitude of the Problem

The demand for wood is expected to increase at least 60 percent during the next 50 years. The growing emphasis on using the nation's forests to help solve energy problems may greatly accelerate this demand. Timber supply will also increase, but at a slower rate than demand. Even considering that projected trends indicate some growth of net imports to the year 2020, there will be an increasing reliance on domestic timber to supply timber products beyond the year 2000. The greatest opportunity to increase the nation's timber supply is on nonfederal forest land, where intensive management treatment on the most productive land can significantly increase production.

While the demand for wood and other products from forest land is increasing, the land base to supply the needs is decreasing. The area of nonfederal forest land has been decreasing in recent decades. Most of the loss occurred as areas were converted to cropland, pastureland, and urban land. These trends are expected to continue.

During the 1980's, a significant portion of the projected decline in nonfederal forest land is expected to result from conversion of forest to cropland. After 1990, however, most of the loss of forest land will probably result from urban expansion and conversion to other uses such as reservoirs, highways, and airports. While the overall trend has been a decrease in the area of nonfederal forest, there can be significant shifts from one land use to another within a short time span (USDA, 1977).

Processes

Productive capacity of many acres of forest land is currently declining for a number of reasons. Good survey data are not available to estimate this acreage, but even a moderate decline of 10 percent over much of the forest land base amounts to large quantities of wood. Numerous soil-site studies on forest land have all shown a definitive relationship between depth of topsoil and forest productivity. Loss of top soil layers through any kind of erosion processes, therefore, generally decreases productivity through loss of both nutrients and water storage. Accelerated erosion on forest land must be viewed as a generally destructive process.

The role of essential elements as a controlling factor in production of forest crops has been demonstrated for all forest areas of the United States in recent years. Many of our forest soils have low or deficient levels of some of these nutrients. Past harvesting regimes entailing long growth periods and high amounts of forest residues left in place encouraged internal mineral cycling to supply essential elements, even on soils with low total supply. Whole tree harvesting in shorter cycles can significantly modify the mineral supply cycles in forest soils as a number of recent studies have pointed out. Reduction in elemental supply from these harvesting procedures can lead to reduction in productivity.

Reduction in essential elemental supply to forest soils can also occur by action of atmospheric inputs which may cause internal leaching. Acid precipitation on already rather poor forest soils is a classic case. However, for some forest areas with sulphur and nitrogen deficiencies, such inputs may increase productivity.

Water deficiency has always been identified as one of the principal factors that reduces forest productivity. Forest practices that lead to overland flow runoff rather than infiltration, and to generally poor use of water, can reduce productivity. Proper nutrition of plants has also been long recognized as a means of making more efficient use of soil moisture.

Much of the reduction in productivity now manifest on forest land originates in low stocking of desirable species. In other words, the energy-fixing capacity of the forest land is not being utilized because the area does not have sufficient tree cover. Correction of this must come from better regeneration practices and better overall management, which in turn comes from better investment opportunities.

These are the primary problems constraining the continued and enhanced production from the nation's forest lands. They are all researchable and some progress has been made. The decade of the 80's must see more definite answers to provide a basis for management and investment decisions.

Research Needs

There is a need to sustain and increase the productivity on forest land to meet future wood fiber demands. Research is needed to bring nonstocked and understocked stands to full productivity; to evaluate the potential decline in productivity caused by erosion of topsoil and nutrient depletion; and to increase the productive capability of the forest land base through improved genetic stock, forest fertilization, manipulation of forest nutrient cycles, i.e., biological nitrogen fixation, and other intensive forest treatment.

The need to reforest stands that are currently nonstocked or understocked will require research to improve regeneration techniques and to match the various forest sites with the proper tree species.

The potential decline in productivity needs to be addressed. The effect of topsoil erosion and intensive harvesting on nutrient depletion and the effect of accelerated nutrient leaching losses caused by acid precipitation will require a major research effort. Considerable opportunities exist for increasing the productivity of the forest land base. Increases in forest productivity have been demonstrated through past research. However, refinements in research are needed to answer such questions as:

1) Where and by what techniques can intensive forest management produce the greatest or most economical wood fiber yields? There is a need to identify those areas where this can occur and devise suitable site-specific treatments for increasing wood fiber yield.

2) How does intensified wood fiber production interact with other forest uses? The impacts of management practices on aesthetics, wildlife, grazing, recreation potential, and other values must be evaluated.

3) Can intensified wood fiber production be achieved without degrading water quality? Research must continue to find better methods and techniques for minimizing disturbances during intensive forest treatments.

Consequences of Implementing Solutions

Implementing solutions to these productivity questions will allow us to become more efficient in capturing solar energy through growth of forest plants for food and fiber. One proven example comes from 10 years of forest fertilization data in the Pacific Northwest. For soils and forest stands that show response to nitrogen fertilizer, at least one extra cubic foot of wood is produced for every pound of nitrogen applied. Energy investments in the production and application of the pound of nitrogen are about 40,000 B.T.U.'s. A cubic foot of Douglas-fir wood represents about 300,000 B.T.U.'s.

This increased production and energy capture will allow us to more fully meet wood supply needs of the nation as well as other forest demands. It is conceivable that we can do this on fewer total acres by concentrating improved practices on forest land more amenable to wood production. This could allow us to make less demands on those forest ecosystems that are more fragile and more difficult to manage.

CONCLUSIONS

Forest land makes up about 30 percent of the total land area in the United States and is the major source of the nation's water supplies. In recent years, the area of forest land has decreased, a trend that will probably continue in the future. At the same time demands are increasing on the remaining forest land for timber, fuel, waste disposal, wildlife, recreation, range, water, and mineral resources. Unfortunately, some of these uses of forest land may not be compatible. Thus, disputes regarding land allocation and uses will increase.

Concerns about the effects of soil erosion on soil productivity on forest land are similar to those for agricultural lands. Off-site or downstream effects—flooding, sedimentation, low flows, pesticide transport, and damage to aquatic habitats—are also similar. However, the wide variety of animal and plant species, and land users, and the extremes in topography, geologic features, and climatic characteristics of many forest areas result in different problems. For example, in contrast to major soil loss by sheet and rill erosion on cultivated lands, most of the soil losses on forested lands are from a relatively small percentage of the area and most of this soil loss is by gullying, mass failure (landslides), and stream channel erosion.

Water problems can be summarized as too much, too little, or too dirty. The volume and timing of streamflow as well as its quality are especially important in forest land management because of their effects on fish and wildlife and on downstream users. Forest productivity is closely linked to erosion and sedimentation processes and to nutrient cycling. Riparian zones are particularly important because of their high productivity, habitat value, and effects on stream properties.

Soil and water research needs on forest land fall into five main areas:

1. Erosion and Sedimentation

Erosion and resulting sedimentation are the most serious soil and water problems resulting from activities on forest land because of their effects on flooding, water development projects, and fish. Erosion and sedimentation processes are different enough from those on agricultural land that research specific to forest land is required in order to meet existing water quality regulation and assure continued producion and use of forest lands. Better techniques are needed to predict the effects of forest-management activities on erosion and sedimentation.

2. Quantity and Timing of Streamflow

It is vital that we understand how forests can be managed to supply the right amount of water at the right time. We know a great deal about the hydrology of forest watersheds. An even better understanding requires more fundamental research on the effects of forest management on: (a) water use by plants, (b) water pathways to streams, and (c) snow accumulation and melt.

3. *Water Chemistry*

Streams from undisturbed forests are usually of high chemical quality. Wildfire, acid precipitation, and forest-management activities such as harvesting, fertilization, prescribed burning, and pesticide applications can have harmful effects on chemical quality of forest streams. We need to know how various natural and introduced chemicals cycle through forest ecosystems and eventually affect the quality of lakes and streams. Such information would aid in selecting practices to maintain or improve the chemical quality of streams.

4. *Effects of Forestry Practices on the Riparian Zone*

Although relatively small in acreage, the riparian zone is a critical ecological land type. Research is needed to understand the impact of forestry practices on water chemistry, temperature, sedimentation, and forest debris in this zone. Understanding these impacts is necessary for selecting proper management practices.

5. *Forest Land Productivity*

Continued or improved levels of productivity on forest land are necessary to meet product and use demands. Principal kinds of research necessary to help obtain this productivity are: (a) research that will enable the site-specific classification of productivity for all forest soils and an understanding of factors that determine productivity; (b) research on factors that could reduce forest productivity, such as nutrient depletion by whole tree harvesting, fire, acid precipitation and erosion; and (c) research on processes and practices that enhance forest soil productivity, such as biological nitrogen fixation, mineral weathering, and fertilization.

FOOTNOTES

[1] Land capable of producing 20 cubic feet of industrial wood per acre per year in natural stands and not withdrawn for other uses.

[2] Organic Act of June 4, 1897 (30 Stat. 35, as supplemented; 16 U.S.C. 475)

[3] Weeks Act of March 1, 1911 (36 Stat. 962, as amended; 16 U.S.C. 515, 521).

[4] Act of June 12, 1960 (74 Stat. 215, as amended; 16 U.S.C. 528-531).

[5] Act of August 17, 1974 (88 Stat. 476, as amended; 16 U.S.C. 1600-1614).

[6] Act of October 22, 1976 (90 Stat. 2949; 16 U.S.C. 472a, 476, 476 (note), 500, 513-516, 521b, 528 (note), 576b, 594-2 (note), 1600 (note), 1600-1602, 1604, 1606, 1608-1614).

[7] Act of August 4, 1954 (68 Stat. 666, as amended; 16 U.S.C. 1001-1008, 33 U.S.C. 701b (note).

[8]Soil and Water Resources Conservation Act of 1977 Public Law 95-192 (91 Stat. 1407; U.S.C. 2001, et. seg. (November 18, 1977).

[9]Act of January 1, 1970 (83 Stat. 852 as amended; 42 U.S.C. 4321, 4331-4225, 4341-4347).

[10]Act of October 18, 1972 (86 Stat. 816, as amended; 33 U.S.C. 1151).

[11]Potential productivity is the amount of wood per acre per year that can be produced in fully stocked stands.

REFERENCES

1. Anderson, Henry W., M.D. Hoover, and K.G. Reinhart. 1979. Forests and water: Effects of forest management on floods, sedimentation, and water supply. USDA For. Serv., Gen. Tech. Rep. PSW-18, Pac. Southwest For. and Range Exp. Stn., Berkeley, Calif.
2. Bengston, George W. 1979. Forest fertilization in the United States: progress and outlook. J. For. 77:222-229.
3. Douglass, J.E. and W.T. Swank. 1972. Streamflow modification through management of eastern forests. USDA For. Serv. Res. Pap. SE-94. Southeastern Forest Exp. Stn., Asheville, N.C. 15p.
4. Duffy, P.D., J.D. Schrieber, D.C. McClurkin, and L.L. McDowell. 1978. Aqueous- and sediment-phase phosphorus yields from five southern pine watersheds. J. Environ. Qual. 7(1):45-50.
5. Hibbert, Alden R. 1979. Managing vegetation to increase flow in the Colorado River Basin. USDA Forest Service Gen. Tech. Rep. RM-66, Rocky Mtn. For. and Range Exp. Stn., Ft. Collins, Colo.
6. Leaf, Charles F. 1975. Watershed management in the Rocky Mountain subalpine zone: The status of our knowledge. USDA For. Serv. Res. Pap. RM-137, 31p. Rocky Mtn. For. and Range Exp. Stn., Ft. Collins, Colo.
7. Likens, G.E., F.H. Bormann, D.W. Fisher, N.M. Johnson, and R.S. Pierce. 1970. Effects of forest cutting and herbicide treatment on nutrient budgets in the Hubbard Brook watershed ecosystem. Ecol. Monogr. 40(1):23-47.
8. McColl, John G. and David F. Grigal. 1979. Nutrient losses in leaching and erosion by intensive forest harvesting. Proc. Impact of Intensive Harvesting on Forest Nutrient Cycling. State Univ. of New York, College of Environmental Sci. and Forestry, School of Forestry, Syracuse, NY, pp. 231-248.
9. Meehan, William R. 1975. Anadromous fish habitat research program for the Western United States. Prelimiary draft, June 6, 1975. Pac. Northwest For. and Range Exp. Stn., Corvallis, Ore.
10. —-, Frederick J. Swanson, and James R. Sedell. 1977. Influences of riparian vegetation on aquatic ecosystems with particular reference to salmonid fishes and their food supply. USDA For. Serv. Gen. Tech. Rep. RM-43, Rocky Mountain Range Exp. Stn., Fort Collins, Colo. p. 137-145.
11. Megahan, Walter F. , William S. Platts, and Bert Kulesza. 1980 Riverbed improves over time: South Fork Salmon. Symp. on Watershed Management, 1980, Vol. 1, Boise, Idaho. p. 380-395.
12. Norris, Logan A. 1978. Toxic materials in forest streams. Proceedings Seminar, 1978, Water Resources Research Institute, Oregon State Univ., Corvallis, Ore. p. 43-68.

13. Pinchot, Gifford. 1947. Breaking new ground. Harcourt Brace Co., Inc., New York.
14. Renfro, G.W. 1975. Use of erosion equations and sediment-delivery ratios for predicting sediment yield. Proc. Sediment Yield Workshop, Oxford, Miss. Nov. 1972, USDA Publ. ARS-S-40, June 1975, p. 33-45.
15. Stone, Earl. 1973. The impact of timber harvest on soils and water. In Report of the President's Advisory Panel on Timber and the Environment. Part III, Append. M, p. 427-467. U.S. Government Printing Off., Washington, D.C.
16. Swank, Wayne T., and James E. Douglass. 1974 Streamflow greatly reduced by converting deciduous hardwood stands to pine. Science 185:857-859.
17. Swanston, Douglas N., and Frederick J. Swanson. 1976. Timber harvesting, mass erosion, and steepland forest geomorphology in the Pacific Northwest. p. 199-221. Donald R. Coates, (ed.), In Geomorphology and Engineering, Dowden, Hutchinson and Ross, Inc., Stroudsburg, Pa.
18. U.S. Environmental Protection Agency. 1973. Processes, procedures, and methods to control pollution resulting from silvicultural activities. EPA 430/9-73-010, Off. of Air and Water Programs, Washington, D.C.
19. U.S. Department of Agriculture, Forest Service. 1980. An assessment of the forest and range land situation in the United States. FS-345.
20. U.S. Department of Agriculture, Forest Service. 1980. An analysis of the timber situation in the United States, 1952-2030.
21. U.S. Department of Agriculture, Soil Conservation Service. 1978. The 1977 national resource inventories. Soil Conservation Service, Washington, D.C.
22. U.S. Department of Agriculture, Soil Conservation Service. 1977. Potential cropland study. Statistical Bulletin No. 578.
23. Wischmeier, W.H., and D.D. Smith. 1965. Predicting rainfall-erosion losses from cropland east of the Rocky Mountains. Handb. No. 282, USDA, Washington, D.C.

7

Donald W. Hedrick and Joseph L. Schuster,
Co-Chairmen

Fee Busby, Carlton H. Herbel, Gordon Kearl,
Henry F. Mayland, Kenneth Renard,
Thomas N. Shiflet, James G. Teer, Dick Whetsell

RANGELAND

Conservation Needs, Technology, and Policy Alternatives

Range—or more appropriately rangeland–is a kind of land and not a use of land. Rangeland provides or is capable of providing forage for grazing or browsing animals. Range includes grasslands and shrublands and those forest lands that continually or periodically, naturally or through management, support an understory of herbaceous or shrubby vegetation that provides forage. Also included are those lands that have been seeded to non-native plants but are managed like lands that support native vegetation. Lands designated as improved pastures, cropland pasture, and grazed croplands are not classified as rangeland because they are routinely cultivated, seeded, fertilized, and/or irrigated.

Rangeland managment is limited by environmental and economic factors. Many rangelands are dry or precipitation is so erratic that arid land management practices must be followed. However, riparian zones that occur as part of the rangeland ecosystem must be managed according to their particular characteristics. Other environmental characteristics that dictate range management options are steep slopes, short growing seasons, and unfavorable soil conditions. All of these environmental factors lower the productive capacity of the land and limit the amount and reqularity of economic input into management. Cultural improvement practices, such as brush management and seeding, are most useful in attempting to improve deteriorated conditions or in meeting special management objectives. Economically inexpensive and energy efficient range management practices, such as controlled grazing and prescribed burning, can then be used to maintain desired ecological conditions.

Rangeland Use

Rangelands are important to the nation. Thirty-nine percent of the land resource base is classified as rangeland. An additional 20 to 30 percent of the land is managed as range. Over 99 percent of this resource occurs in the 17 contiguous Western States and Alaska, but rangeland must not be thought of as a Western resource. Rangeland is important to all Americans.

Rangeland provides forage, cover, water, and space for domestic livestock and wildlife. Rangeland or lands used as range provide over 200 million animal-unit-months of domestic livestock forage. The sheep and goat industry in particular depends upon rangeland grazing; 80 percent of all sheep and goat products,including lamb, mutton, and goat for human consumption and wool and mohair for clothing, are produced on rangeland. Rangeland is the most important habitat for wildlife in the United States, and healthy wildlife populations depend upon how well we manage these lands.

Water originating from rangeland is an important part of the water supply for the nation.

People use rangeland directly for a variety of recreational opportunities, taking advantage of the clean air, open space, and healthy, natural environment that rangeland provides. The role that these get-away-from-it-all activities play in the social well-being of the nation is unquantified but certainly important.

Rangeland is a source of energy and mineral supply. Coal, oil and gas, oil shale, uranium, and numerous other materials occur on or under rangeland. These materials are being "harvested" to meet our energy and mineral requirements. Lands disturbed for such purposes must be reclaimed to insure their usefulness in meeting future demands of society.

Multiple use is generally practiced on rangeland, but some of the uses may influence management decisions more than others. For instance, an area of range may be used for livestock grazing, wildlife habitat, and water supply, but one of these uses may be of higher priority today because of society's demands or land ownership desires.

Management Objectives

Management objectives for rangeland vary among areas depending upon production capabilities and socioeconomic demands. A primary objective for managing rangeland is to provide a mix of products, services, and values demanded by society today without reducing future production and use options through deterioration of soil and water resources.

Use of rangeland for production of food for human consumption will become increasingly important in the future. A growing population, both in the United States and worldwide, will require that some grain now used to feed livestock be diverted for human consumption. In addition, some grain now fed to livestock may be used for conversion to energy fuels such, as alcohol. Range forage can be converted to human food by the grazing animal without large inputs of costly fossil fuel. Range-fed livestock will become an even more important source of meat for humans. Thus, rangeland will contribute to maintaining a quality diet for Americans.

CURRENT SITUATION

Misuse of western rangeland in the 19th century resulted in a deteriorated ecological condition and reduced forage productivity. Although management of rangeland under federal control improved in the early 1900s with the establishment of national forests, most of the public domain was not brought under management until 1936. Effective management programs really began after World War II when the Grazing Service, responsible for managing these areas, officially became the Bureau of Land Management.

Emphasis on improved management of nonfederal rangeland started in the mid 1930s with creation of the Soil Conservation Service and efforts by the land grant institutions through extension and education programs. Limited data exist describing the condition of nonfederal rangeland at that time. Table 1 contains estimates of range condition by the Soil Conservation Service (SCS) in 1963 and again in 1977. These data indicate that during this 14-year period 20 percent of the nonfederal rangeland in fair or poor condition improved to good or better condition. However, 60 percent of this rangeland still remains in less than good condition (Pendleton, 1979).

Box (1978) reported the percentage of all federal land in three condition classes and for three dates starting with data from Senate Document 199 (U.S. Senate, 1936). Marked change in condition was apparent between 1936 and 1966, but little change occurred between 1966 and 1972.

Although data from various sources may differ from the trends in Tables 1 and 2, there appears to be general agreement that the trends are representative of changes taking place on Western rangelands since the 1930s.

Table 3 summarizes, by major ecosystems, the condition classes as well as the present and potential stocking rates on both public and private land in the 17 Western states. The present stocking rate of 195 million animal-unit-months (AUM's) could be increased to 620 million AUM's if

Table 1. Range conditions, 1963 and 1967.

	Range condition (percentage of area)			
Date	Excellent	Good	Fair	Poor
1963	5	15	40	40
1977	12	28	42	18

Source: SCS, USDA

Table 2. Range conditions, 1936, 1966, and 1972.

	Range condition (percentage of area)		
Date	Good or excellent	Fair	Poor or very poor
1936	16	26	58
1966	18	49	33
1972	18	50	32

Source: Box (1978)

all the land were in good condition. Interestingly, the data indicate that 15 percent of the range is in good condition; 31 percent in fair; with the remaining 54 percent being in poor or very poor condition.

Frequent drought, coupled with encroachment of unwanted plants and past management practices, has resulted in the deterioration of much western rangeland. Brush has replaced much of the grass and other forage on 200 million acres of rangeland in the Southwestern alone (NACD, 1979). This leads to reduced livestock grazing capacity, inefficient use of precipitation, increased soil erosion, and deterioration of wildlife habitat. Seventy-seven percent of all nonfederal rangeland needs some kind of conservation treatment.

Sheet and rill erosion on rangeland ranged from 0.5 to 8 tons per acre per year in 1977. The average soil loss was 2.8 tons. Rangeland erosion is a severe problem (over 5 tons/acre/year) in two range states, a major problem (3 to 4.9 tons/acre/year) in one state, and a moderate problem (2 to 2.9 tons/acre/year) in seven states. Average rangeland erosion is less than 2 tons per acre in seven of the 17 Western states.

Because many rangeland soils are shallow and have considerable slope and climatic limitations, erosion occurs readily where land is misused. Wind erosion in excess of 2 tons per acre per year occurs on 27 million acres of rangeland in the 11 Great Plains states. This represents about 9 percent of that regions rangeland. Sheet and rill erosion on rangeland in excess of 2 tons per acre occurs on 124 million acres or 30 percent of all nonfederal rangeland, excluding Alaska. States with severe sheet and rill erosion problems on rangeland include Arkansas, California, Colorado, and Mississippi. Texas also has a major sheet and rill erosion problem with over three tons per acre loss. Moderate sheet and rill erosion is found in Kansas, Nebraska, Nevada, New Mexico, Oklahoma, Utah, and Wyoming. Most other range states have only slight erosion problems (USDA, 1980b, c). Wind erosion is a problem in New Mexico and Texas.

The reduced carrying capacity of range in less than excellent condition represents a loss of 30 to 40 million AUM's of grazing, or a loss of

Table 3. Rangeland types, areas, condition, and stocking in the 17 Western States.†

Total stocking ecosystem‡	Area	Condition				Grazed	Stocking rate	Present	Potential
		Good	Fair	Poor	Very poor				
	million acres	%					AUM/a	1,000 AUM's	
Sagebrush	129.9	12.3	36.1	35.0	16.6	89	0.182	24,641	77,770
Desert shrub	81.2	17.4	36.4	31.5	14.7	70	0.041	2,809	8,070
Southwestern shrubsteppe	43.2	10.5	14.8	41.5	33.2	92	0.084	3,775	17,140
Chapparral-mountain shrub	15.5	11.8	24.0	37.9	26.3	95	0.125	1,612	8,300
Pinyon-juniper	47.3	9.1	28.6	44.4	17.9	88	0.073	2,393	12,040
Mountain grasslands	26.9	17.5	36.2	31.2	15.1	97	0.319	16,597	29,020
Mountain meadows	3.3	32.4	38.8	21.1	7.7	79	0.613	822	4,220
Desert grasslands	24.7	8.4	24.7	50.5	16.4	93	0.091	2,998	8,220
Annual grasslands	10.2	13.2	5.0	36.7	45.1	96	1.095	10,649	64,010
Alpine	6.8	42.8	27.1	28.7	1.4	70	0.045	216	460
Shinnery	4.7	16.5	27.7	40.5	15.3	97	0.406	1,848	6,780
Texas savanna	28.4	16.4	23.5	45.9	14.2	99	0.428	16,493	43,660
Plains grasslands	175.2	14.7	34.2	39.6	11.5	98	0.294	54,325	168,560
Prairie	41.0	14.0	34.2	38.1	13.7	95	1.090	45,145	148,680
Douglas-fir	38.5	--	--	--	--	52	0.049	1,000	2,000
Ponderosa pine	33.7	--	--	--	--	74	0.059	1,617	3,230
Western white pine	0.6	--	--	--	--	48	0.234	50	100
Fir-spruce	113.4	--	--	--	--	51	0.028	418	840
Hemlock-Sitka spruce	19.8	--	--	--	--	0	0.000	173	350
Larch	2.8	--	--	--	--	41	0.147	178	360
Lodgepole pine	21.2	--	--	--	--	56	0.011	527	1,050
Redwood	0.8	--	--	--	--	55	0.063	28	60
Hardwoods	39.8	--	--	--	--	56	0.059	1,222	2,440
Longleaf slash pine	0.3	--	--	--	--	96	0.200	55	110
Loblolly-shortleaf pine	5.4	--	--	--	--	37	0.011	22	40
Oak-pine	3.3	--	--	--	--	48	0.105	147	290
Oak-hickory	7.6	--	--	--	--	28	0.253	531	1,060
Oak-gum-cypress	2.8	--	--	--	--	3	2.490	191	380
Elm-ash-cottonwood	1.6	--	--	--	--	10	0.213	51	100
Wet grasslands	3.5	17.0	27.6	20.7	34.7	79	0.663	4,304	8,610
Desert	7.5	76.6	11.6	6.5	5.4	45	0.001	3	10
Total	940.9							194,840	617,960

† Partially obtained from and based on "An Assessment of the Forest and Range Land Situation in the United States" (Forest Service FS-345, 1980).

‡ Potential stocking was calculated by assuming that good, fair, poor, and very poor range conditions produced 80, 50, 30, and 10% respectively, of potential. Based on the ecosystems for which we had condition data, we assumed that the potential stocking for the remaining ecosystems could be approximately doubled.

almost one billion pounds of red meat and significant amounts of wool, mohair, leather, tallow, insulin, and many other by-products. With the increase in demand for red meat and animal products by the year 2030, demand for range grazing is expected to increase 140 percent over what it was in 1976 (USDA, 1980b, c).

A 1973 inventory of brush encroachment problems showed 277 million acres of brush on nonfederal rangeland and pasture. Twenty-two areas were identified as having major brush problems. In these areas brush was so dense it dominated the plant community, seriously suppressed the growth of more useful plants, severely restricted land use, and left soil vulnerable to erosion. These areas contained 223 different kinds of brush. In the 17 Western States severe brush problems were found in Oklahoma, Oregon, Utah, and Texas. Five range states had major problems, four states had moderate brush problems, and four had only slight brush problems.

Land use conversion will continue to reduce the rangeland base. The 1975 Potential Cropland Study (USDA, 1977) indicated that from 1967 to 1975 just over 4 million acres of rangeland and pasture were converted to urban, built-up, and water areas. This equals one-half million acres per year. These figures are not great nationally, but locally they are significant. Future losses in the rangeland base will occur chiefly through conversion to urban, built-up, and cropland uses. The SCS Potential Cropland Study shows that about 40 million acres of rangeland has medium potential for conversion to cropland, while 9 million acres has high potential. The RPA assessment projects a 57 million acre loss to all uses by the year 2030.

In the 1980 update of the Forest Service RPA document livestock grazing use in the National Forest Systemis projected to increase from the current 9.9 million AUM's to 15.5 million AUM's by the year 2025. This increase will be made possible by improving coordination with other resource uses. Two key elements in achieving this goal are (1) grazing in timber stands for 10 to 20 years after harvesting and (2) concentrating grazing on the most productive lands.

STATEMENT OF THE PROBLEMS

The Problems

The nation is faced with an almost continual reminder of the need for increased levels of production and quality of food, fiber, and environmental amenities. Recently, this plea has been muted by the supreme dominance of energy over our economy, politics and society. Concomitant with the concern about energy is an increasing concern for natural resources, namely, soil, water, and air. This situation emphasizes that we are finally in a resource-oriented era.

The worldwide population increase is projected at 1.7 percent annually. At this rate there will be 3 to 6 billion more people by the year 2030. In the United States an increase from about 230 to 300 million is expected during the same period. The rate of increase in food and fiber production has been about 1 percent a year. Much has been written about whether the world will have an adequate food supply in the future when 15 million people already die annually from starvation.

Among the constraints facing the United States and the world are the following:

1. Shortage of fossil fuel energy.
2. Scarcity of water.
3. Insufficient knowledge and technology reserve.
4. Failure to apply existing technology.
5. Increasing competitive uses for resources.

As the demand for goods and services from rangelands increases, these constraints make future range management qualitatively different from that practiced in the past.

By its very nature, rangeland is managed for a variety of products. This attribute is the basis for conflict among several segments of American society, and in some cases it has polarized opinion to the detriment of rangeland management for any of the products.

The dominant or primary use of rangeland resources in the past 100 years on public and private rangelands has been the production of domestic livestock for meat, wool, and mohair. Current users of these rangeland resources are now more numerous, affluent, environmentally concerned, and have other values than pre-World War II society. Water, hunting and fishing, nonconsumptive recreational pursuits, oil, gas, hard rock minerals, timber for wood and fuel, and even scientific and aesthetic values are among those valuable products that can be produced and are being requested by Americans. These products have value in the marketplace.

Research is needed to find ways of integrating management for as many rangeland products as possible. In some land forms, where terrain, remoteness, climate, and soil and water resources limit production and/or uses of several products, a single use may be an essential management goal. On most range sites, however, more than one product can be produced without compromising or reducing the production of others. Indeed, many extensive rangelands, grazing by domestic livestock is the most effective tool for producing vegetative conditions favorable to other uses.

Society's demands may require single use or multiple use. These decisions, nonetheless, should have the guidance of scientific information.

Although progress has been significant in improving the productiveness of rangeland in the past three decades, much remains to be done. Several problems are unique to rangeland and its proper management. They include:

1. Range management decisions are difficult because of the lack of

knowledge about the interdependence among the multiple uses of range ecosystems and lack of climatic and economic data and analysis.

2. Competing biologic components reduce productivity.

3. Water shortages limit productivity.

4. Knowledge of nutrient cycling in relation to soil-plant interactions, soil fertility, water supply, and animal performance and production is inadequate.

5. Knowledge of optimum livestock breeds and cross breeds, and wildlife species for various ecosystems and management systems is lacking.

In the future most cropland suitable for feed and fiber crops will likely be used to produce food and other products for human consumption. Moreover, energy shortages call for better management of our natural resources.

Demands for recreation, watershed protection, and fossil fuel mining further increase the need for research. Also, range improvements enhance wildlife habitat and promote soil and water conservation. The needed conservation treatment of water developments, fencing, brush and noxious plant control, vegetation type conversion, and revegetation of ranges in poor condition requires time and money.

The Solutions

Problems facing range in the next three decades can be partially met by application of known technologies. Much can be done if what is known can be applied through education and extension. However, the knowledge base is not complete, and basic information is needed on an array of problems before the nation's ranges will be able to meet the demands put on them by 2030.

The decade of the 1970s brought out many projections of range research needs. These reports identify an array of researchable problems. Invariably, some generally inclusive research needs emerge:

I. Improvement, maintenance, and use of rangelands for efficient utilization of soil and water resources

Previous use and management of rangeland, along with climatic, soil, and other environmental conditions, have resulted in losses and damage to rangeland soils. Water quality, and quantity and timing of water yields have also been affected adversely. Rangeland now produces forage at less than half its potential. Frequent drought, rodents, insects, and brush encroachment are major causes and indications of deteriorated conditions. Major gains can be realized from research aimed at rehabilitating deteriorated rangeland. Restoration of range ecosystems to correspond more closely with their ecological potential should improve stability of all range resource values as well as increase forage supplies.

A. Range Improvement Practices

1. Brush Management. Brush is a problem on much of the

approximately 940 million acres of land grazed in the United States. Brush on rangeland induces losses by decreasing forage production, water yield from watersheds, and recreational activity (hunting, fishing, picnicking, etc.). Dense stands of poisonous, thorny, and pollen-producing brush species also increase the cost of handling livestock and losses of livestock through death and injury. Human allergies are aggravated as well.

Methods of controlling many brush species on rangeland are ineffective, inadequate, or not economical. Chemical application, a common practice, provides temporary control. Treatments must be repeated in subsequent years. Mechanical treatments, such as dozing and root-plowing, may effectively control brush, but they require large energy inputs, sometimes destroy the grass stand, and expose the soil to wind and water erosion. Other mechanical treatments, such as mowing, chaining, and roller chopping, require less energy input and soil disturbance but usually provide only temporary control.

Methods to determine accurately the losses caused by a given infestation of brush would provide a useful indicator of when control is economically justified. Accurate data on brush control costs and environmental effects of each control method would provide a basis to judge the value and feasibility of alternative methods.

The research objectives are to develop new and improved methods of brush control that are safe, effective, and economical for the major problem species on rangeland.

2. Revegetation. Millions of acres of rangeland in the United States have been and are being disturbed by plowing, mining, road construction, drought, and poorly managed grazing. After severe disturbances, annual plant species are first to re-colonize the site. Then come pioneering perennial species that, in turn, may be followed by the perennial dominant species. Unfortunately, the procession of species sometimes requires an unbearably long time and is sometimes prevented entirely by soil losses or other irreversible changes in the ecosystem.

For this reason, an urgent research need is the development of technology required to obtain prompt revegetation of disturbed areas for conservation of soil resources and livestock production. Methods of revegetation can be defined simply as sodding or seeding; however, the technology must be based on knowledge of soils, climate, range sites, plant species adapted to those sites, mechanical and chemical methods of seedbed preparation, sodding methods and equipment, methods and seasons of seeding, methods of conserving soil moisture, seed and seedling characteristics and requirements, and management requirements after seeding.

The real challenge is to define the combination of treatments and methods that minimize effort and cost on each range site. Some results transfer from one site to another, but the limitations of methods and techniques can be defined only by repetition on many different sites. Such repetition must not be misconstrued as needless duplication of research effort.

The major objective of this research is to develop knowledge of species, soils, microclimates, and revegetation techniques that, when properly combined, will establish useful plants on depleted ranges. The species should promote conservation of soil and water resources.

B. Grazing Systems Research

Grazing management is the most important single practice in ranch and range management. It brings together the elements of range improvement into an overall system for production of meat, wool, and mohair. It applies to every acre of grazed land. All other range improvement practices fail if grazing management fails. However, without seeding, brush control, insect and rodent control, and other practices where and when they are needed, grazing management alone could never bring a range to its maximum value to society.

Effective use of diverse range resources will require development of management systems compatible with many complex biological factors. Native range responds to grazing intensity, season of grazing, length of grazing season, and the interaction of grazing management, soil, and climatic factors. At the same time, animal productivity changes as the grazing animal matures and responds to environmental factors, including quantity and quality of forage. Intensive research is needed to provide basic information on the response of each plant and animal species to intraspecies and interspecies competition and to provide guides to the interaction of the biological systems involved in the soil-plant-animal complex. With this basic information, years can be saved in developing and testing grazing systems.

The research objective is to develop improved production systems for converting range forage to livestock products, consistent with conservation of soil and water resources.

C. Improved Plants and Animals for Efficient Use of Soil and Water Resources

Improvements in the genetic ability of range plants to produce quality forage is basic to improving rangeland productivity. Plant breeders can improve the genetic ability of a species to produce a useable product only if genetic variability exists within the species. Germplasm of native and introduced range plants must be collected, evaluated, and preserved in genetic banks for range plant breeders.

Range plants include many diverse species with unique reproductive systems and complex cytology. Often they have minute flowers that discourage emasculation and hybridization. Apomixis, cleistogamy, polyploidy, chromosomal irregularities, and taxonomic problems are common and limit the use of conventional breeding methods in many valuable range species. Cytogenetic and cytotaxonomic studies are needed to solve reproductive problems, establish species relationships, and develop breeding techniques.

Breeding of cultivars for specific sites and special uses, such as wildlife habitat, conservation, reclamation, and enhancement of environmental quality, is also needed.

The research objective is to produce genetically superior forage plants for livestock production, wildlife uses, and specialized purposes on major range sites and localized problem sites.

The feed efficiency in beef cattle, sheep, and other animals is enhanced by their ability to use roughages as a source of energy and non-protein nitrogen as the major source of dietary protein. These attributes have not been fully realized.

Increased reproductive efficiency per animal unit would greatly improve the production of meat per acre of rangeland. Average reproductive efficiency in American beef herds and sheep flocks has been increasing. However, progress is much slower than what could be realized through increased research on those factors affecting the reproductive performance of range livestock and wildlife.

The necessary methods and systems for increasing biological efficiency in animals maintained chiefly on rangeland could be identified and perfected through integrated, interdisciplinary research.

Research objectives are to develop livestock with high biological efficiency for meat production and to use range animals as a means of maximizing efficient use of rangeland resources with minimal use of fossil energy.

II. Develop an inventory classification system applicable to rangeland for maintaining and improving the use of soil and water resources

A. Research is needed to determine the relationship between soil and water losses from rangelands and the continued productivity of the range resource.

Many attempts to apply present technology to improve productivity are complicated by the extreme variability of factors affecting these fragile sites. Nevertheless, a national uniform system of resource inventory and classification is needed that will permit documentation or evaluation of the impacts of methods to improve range condition. Such a system would relate chemical and physical properties of soil to infiltration, plant growth, soil moisture, evapotranspiration, and erosion.

B. Allowable soil loss values (T) need to be developed for rangelands

Information does not exist that allows the evaluation of soil erosion effects on changes in productivity. Soil loss from rangeland is highly variable because of climatic, physiographic, and edaphic characteristics. With an already restricted soil profile, any erosion losses may be intolerable. Research is needed to define tolerable soil loss. The rate of soil formation from parent geologic material offsets the erosion loss, and the sum must ensure long term soil productivity. Existing, limited information suggests that the natural soil-forming processes are much slower under semiarid conditions than under more humid conditions. Low T-values are much more critical on shallow soils than on deep soils. Such fundamental research must include plant nutritional and water requirements, consideration of fundamental erosion mechanics from both wind and water, and the consequences of erosion for on-site fertility and downstream water quantity and quality.

C. Uniform range inventory procedures, including use, suitability, and classification of range potential, are needed.

These procedures would allow more precise application of the best management practices required by legislation for the conservation of soil and water resources. Such programs are also needed to ensure the quality of downstream water resources.

SUMMARY

Range includes grasslands, savannas, shrublands, and forests that can be used by grazing or browsing animals. Economic and environmental factors such as low productivity, little precipitation, steep slopes, and shallow soils limit how rangeland can be managed.

Nearly two-thirds of the land in the United States is used as range. This land is a source of forage, cover, water, and space for domestic livestock and wildlife. Rangeland provides more than 200 million animal-unit-months of grazing for livestock. It also represents some of the most important wildlife habitat in the nation. Rangeland provides a variety of other commodities, and uses as well, including recreation, water supply, and energy and mineral resources.

Although the idea of multiple use generally applies to rangeland, some uses may influence management decisions more than others. A primary objective of range management is to provide the best mix of products, uses, and values demanded by people today without reducing future production and possible uses through land deterioration.

Because of competition from other uses for grains now fed to livestock and because of conversion of crop and pasture land to other uses, we expect more of the nation's meat supply to be produced on rangeland. The ability of grazing animals to convert range forage to human food means that rangeland will help maintain a quality diet for Americans.

Researchers, users, and policymakers must find ways to improve any rangeland whose potential to meet future demands is threatened because of inadequate management. Two problem areas deserve research if we are to conserve soil and water resources on the nation's rangeland:

1. Rehabilitation and Management of Rangeland for Efficient Use of Soil and Water Resources

a. Learning how to rehabilitate deteriorated rangeland should permit restoration efforts that will reduce soil loss, improve water use, and increase productivity. Better and more efficient methods of brush and weed control, revegetation, management of insects, and improved cultural techniques are needed.

b. Grazing systems must be developed that enable livestock to more efficiently use rangeland forage while improving or maintaining soil, water, and vegetation resources.

c. Range plants and animals with greater biological efficiency must be selected and bred. Measures of efficiency include fixation of atmospheric nitrogen, resistance to pests, photosynthetic rate, and the conversion of plant material to animal products with less water use and soil loss.

2. Inventory and Classification of Rangeland for Monitoring Soil and Water Resources

a. We need to learn how to evaluate the effects of soil erosion and water loss on range productivity. This information is needed to make better decisions about range management.

b. We must learn how much soil erosion rangeland can endure before its pro ductivity is adversely affected. Without this information, range managers cannot identify those sites where soil is being lost at a rate that threatens the future use of those sites.

c. Uniform procedures are needed to make inventories of range use, suitability, condition, and trends; and better procedures are needed to classify range potential. These procedures would permit more precise application of the best management practices to conserve soil and water resources.

REFERENCES

1. Box, Thadis W. 1978. The arid lands revisited-one hundred years since John Wesley Powell, Fifty-seventh annual honor lecture at Utah State University, Logan.
2. National Association of Conservation Districts (NACD). 1979. Pasture and range improvement report, Washington, D.C.

3. Pendleton, Donald T. 1979. Extent and condition of nonfederal rangelands of the United States. p. 25-28. In Proc. Symposium on Rangeland Policies for the Future, University of Arizona, Tucson, Ariz.
4. U.S. Department of Agriculture. 1977a. Potential cropland study. Soil Conservation Service. Stat. Bull. 578.
5. —-. 1980b. RPA - An assessment of the forest and rangeland situation in the U.S. Forest Service Rep. FS-345.
6. —-. 1980c. RPA - A recommended renewable resources program - 1980. Update, U.S. Forest Service Rep. FS-346
7. —-. 1980d. RCA Summary of appraisal, Parts I and II, and Program report, Soil Conservation Service.
8. U.S. Senate. 1936. The western range. Senate Document 199. 74th Congress, 2nd Session U.S. Government Printing Office, Washington, D.C.

8

Marvin T. Beatty and Fred P. Miller, Co-Chairmen

W. D. Anderson, Richard L. Barrows,
John Konrad, William H. Langer, Donald Last,
Larry C. Merriam, James C. Patterson,
Harry B. Pionke

URBAN AND RECREATIONAL LANDS

Conservation Needs, Technology, and Policy Alternatives

All life is dependent upon stocks and flows of nutrients. The emergence and development of primeval civilization was largely paced by the emergence and development of agriculture. Soils, therefore, provided the medium through which these nutrient flows and energy conversions could take place. Prior to this development, these nutrient and energy flows were attained by a nomadic hunting and gathering existence. This transition to a conscious plant and animal husbandry predates recorded history, but was surely one of the great forward steps of humanity. Subsequent use of animal power and eventually the industrial revolution itself released large portions of the populace from agrarian ties.

Migrations to the cities began in earnest. Man's natural and cultural environment changed drastically as a result of these population shifts.

Increased population need not compromise the environmental integrity of an area. Ever since Neolithic times, humans have lived in managed environments. Dubos (1976) perceives the earth as "neither an ecosystem to be preserved unchanged nor as a quarry to be exploited for selfish and short-range economic reasons, but as a garden to be cultivated for the development of the human adventure." This "garden" can be cultivated in urban and rural environs and can be nurtured by being left undisturbed in wilderness environs.

Barlowe (1975) states that World War I marks the approximate date when the U.S. shifted from its earlier status as a predominately rural nation to its present status as an urban-oriented nation. Today, as in the past, most of the U.S.A. land resource base is devoted to forests, crops, pasture, and range. These "agricultural" lands account for roughly seven-eighths of our land area. The remaining portion of land is devoted to "special uses" consisting of mining, recreation, transportation, commercial, industrial, and urban uses.

In 1920, when 51.2 percent of the nation's 106 million people were listed as urban and 29.9 percent were classified as rural farm people, only 39 million acres were reported by the U.S. Department of Agriculture as "special use" lands (Barlowe,1975). In 1970, nearly three-fourths of the nation's 203 million people were classed as urban with only 4.8 percent classed as rural farm people. During this half century (1920-1970), the area devoted to "special uses" had increased to just under 170 million acres (Barlowe, 1975). In this half century period (1920-1970), while the U.S. population had nearly doubled, the "special use" land requirements had increased more than four-fold.

Precise data on the conversion of rural land to agricultural use are not available. Soil Conservation Service surveys in 1967 and 1975 indicate that about 3 million acres of rural land are converted to urban built-up or water use annually, and about 70 percent of the total is converted to urban use (Dideriksen, 1977). However, these data are based on a survey methodology rather than a census, and also do not include land that is effectively removed from agriculture, even though not converted, because it is surrounded by urban uses or is not feasible to farm for other reasons.

Future trends on the conversion of rural land to urban use are even less clear. In the past two years, high interest rates and the general economic slump in the housing industry have slowed the rate of conversion of rural land to urban use. However, short-run fluctuations in activity must be separated from long-run trends. Increasing energy prices will increase transportation costs and will tend to discourage dispersed development. In addition, the effect of the post World War II baby boom on housing construction and land conversion may be largely past, which will tend to reduce the rate of conversion. On the other hand, the increase in the number of one- and two-person households, if continued, will tend to increase the rate of conversion. Second, the dispersal of job opportunities into more rural areas, the "population turnaround" in many rural counties, and the preference of many Americans for a rural housing site and lifestyle suggest that rural land will continue to be converted to urban use.

The 89 million acres of state and national parks, wilderness areas and wildlife areas in 1969 account for only a portion of the land area used for recreation in the USA. Approximately 15 percent of the developed land area of U.S. cities is used for parks and playgrounds according to Barlowe (1975).

The Bureau of Outdoor Recreation reports (USDI, 1973) that 319 million acres of publicly owned lands were administered primarily for outdoor recreation uses in 1972. About eight percent (26 million acres) of this total was used for intensive recreational activitiy in local and regional areas.

The way the land is used affects both immediate and remote environments. Runoff from the land, whether from cropland recently treated with fertilizers and pesticides or from urban and industrial areas, affects the water quality of receiving streams. Discharges of auto exhaust and industrial smoke and gases to the atmosphere are dispersed in accordance with climatic patterns and topographic features. The concentration and impact of the fall-out from these discharges, whether in the form of particulate matter, "acid precipitation," or other forms, affect surrounding and remote ecosystems. The character of the vegetation and soils receiving these runoff and atmospheric loads determines in large part the intensity and dimensions of the environmental impact.

Soils and landscapes vary enormously in their capacity to accommodate the variety of uses to which we subject them. Building upon the fragile coastal zones impacts the aquatic web of life and causes a roulette-like invitation for potential hurricane disasters. Building on soils and substrata susceptible to load failure and uneven consolidation causes billions of dollars in structural damage each year plus untold heartache. Soils and landscapes with near-surface watertables and poor drainage characteristics require special management for crop production as well as urban uses. Flooding, water moving into and around basements and foundations, and water-induced vegetation damage are a constant bane for many urbanites.

Homeowners trying to use a variety of soils as a medium for on-site domestic waste disposal soon find out that all soils are not created alike. Local health departments are under relentless pressure to allow installation of on-site waste disposal systems. Then they are besieged to "do something" once these systems fail. Erosion and sediment discharges threaten crop production and cause tremendous environmental impacts in aquatic ecosystems. Sediment impacts stem from both agricultural land uses and many nonagricultural activities such as construction. The chemical and mineralogical attributes of many soils are manifested in many ways from pipeline corrosion to damage of landscape plantings. Other land-use problems include the encroachment of urbanization and "buckshot" development onto prime agricultural land, building in flood plains and earthquake-prone zones, and encumbering or blocking scenic vistas.

These soil and land-use problems are caused by increased population pressure on the land resource. As mentioned previously, not only have the numbers of people increased, but their affluence and mobility have resulted in demand for accessibility to more land. Since 1945, this population increase in the USA has required new housing for more than 70 million people plus the housing needed for internal relocations. To service this added growth, land was needed for our 20th century amenities such as roads, airports, shopping centers, parks, sewer systems, water supplies, schools, police and fire protection, landfills, mineral excavations, industry, fuel processing and storage, electric generating stations, and transmission lines.

In essence, the USA built the equivalent of a nation with the population of France on top of and alongside of what existed at the close of World War II. The encroachment of this growth onto land previously occupied by crops, forests, range, and pasture, and the use of fragile ecosystems and soils unsuited for many of these uses have precipitated severe environmental and land-use problems in many areas. Furthermore, the proximity of many land users to each other and to sensitive ecosystems (e.g. rivers, reservoirs, lakes, estuaries, marshes, wetlands, airsheds) result in unintended and unanticipated impacts and spillovers of land-use by-products. These spillovers and impacts take the form of pollution from runoff, groundwater contamination, smoke, exhausts, odors, noise, and visual blight. Other impacts include increased social tensions, increased traffic congestion, increased taxes, increased health risks and medical expenses, increased resource consumption, reduced efficiency, increased costs of resources, and forfeited options for the future due to the irreversibility of many of these uses.

Governments, especially local governments, are asked to intervene in these problems and provide a solution in the interest of the public welfare. Despite the challenges that these problems pose to local governments, there is often a lack of sound data or even schematic models to assess the land's carrying capacity and to anticipate urban and recreation impacts on the environment.

WATER

A. Urban Water Quantity

1. Current Situation

Water supplies in the United States are adequate to meet present demand except during extreme drought events. A combination of integrated surface-water use and groundwater extraction plus interbasin water transfer can provide all major urban areas with sufficient water through the year 2000. Chronic water shortages presently exist and will continue in small villages and cities that are unable to generate the capital needed to import water or develop local water supplies adequately.

The Rocky Mountains, Great Basin, Great Plains, and High Plains areas are at the brink of chronic water shortages. These areas are marginal since interbasin water transfer is presently uneconomical. Surface water is minimal due to low rainfall, and groundwater is either allocated or in the process of serious depletion by water use in excess of recharge.

Some cities east of the Mississippi River experience water shortages due to drought or damage from flood. Since the East is largely a surface-water user, the groundwater remains a nearly untapped reservoir. Local shortages are mostly due to a lack of adequate reservoir storage and a tendency to discount groundwater as a viable water supply.

The major water problem for urban areas is the periodic shortages that occur in periods of drought or low flow. These shortages are compounded if high urban demands for water occur when supplies are limited.

Since forest and agricultural lands generally occupy about 90 percent of urban watersheds, water and soil conservation practices on these lands are eminently important to the urban water supply in both quantity and quality.

Many urban areas are already facing water shortages, especially during droughts or in summer. Moreover, urban and industrial water use through the USA is increasing both due to increasing population and increasing per capita use. In many areas, municipalities and industries have or can afford to establish use priorities over agriculture or recreation. Projections for the Susquehanna River Basin (New York, Pennsylvania, Maryland) and other basins in the Northeast vary, but generally conclude that municipal and industrial water use will increase 50-100 percent by the year 2000 and 100-200 percent by the year 2020 (Susquehanna River Basin Study Coordinating Committee, 1970, North Atlantic Regional Water Resources Study Coordinating Committee, 1972). Consumptive rural water-uses (including irrigation) in this area are projected to increase similarly and will be in direct competition for runoff and groundwater supplies. In the less water-rich regions of the USA, where additional water supply development is even less likely, this competition will become increasingly more severe.

Urban areas have been shown to increase flood peaks and decrease groundwater recharge. The increased flood peaks increase the severity of flooding on both urban and nonurban areas. Decreased groundwater recharge, which is often aggravated by groundwater withdrawals, causes a declining water table. Depending on properties of the site and location, the impact of this decline may include: 1) subsidence, 2) salt water intrusion for coastal cities, 3) loss of emergency or standby wells for water supply or fire protection, and 4) the loss of wells or pumping capacity for adjacent farm or rural wells. Power demands by urban and other areas require explicit or implicit commitment of water to the power plant throughout the year. The maintenance of a designated minimum flow in streams for power, environmental, or other uses may be a most crucial water-quantity problem in urban areas.

The increasing urban competition for water supplies, and issues such as water disposal of wastes, runoff, and groundwater use and protection affect both city and rural residents. Development and application of the appropriate water conservation technologies both upstream and in the urban areas are needed if the projected trends continue.

2. Statement of the Problem

The increasing urban water needs mostly will be met by additional water-supply development or the acquisition of nonurban water rights. Even in the Northeast where the overall water-supply-development potential is very large, the distribution of developable water resources within watersheds and sub-basins varies greatly, causing some watersheds to have little water reserve. Moreover, water supply in summer or drought periods is often insufficient. In the drier areas of the country, urban water may be continually in short supply and water deficits are often made up diverting waters from other basins. Water conservation methods for urban and rural areas need to be developed and applied. A more efficient water use and control will also improve water quality in many cases.

The need is to increase urban water-use efficiency, control urban effluent and peak runoff rates, and store or retain stored urban water on-site. Advances in urban water-use efficiency may provide more waters locally and areally for nonurban use. The control of peak runoff rates from impervious surfaces and groundwater recharge may reduce in-city and downstream flooding hazards as well as maintain groundwater levels. In addition to the advantages described in the Current Situation, these recharged groundwater zones may serve as supplemental urban water sources during summer or drought periods. Fulfilling these needs may require: 1) joint, comprehensive water-resources development and use of planning that includes urban, nonurban and other impacted users in the area or watershed, 2) application of presently available water-conservation methods and 3.) development of water-conservation methods with particular emphasis on groundwater recharge and the reduction of peak-flow runoff rates. Research into economic and institutional (1), extension and implementation (2), and the controlling physical process, with emphasis on the development of new conservation methods, would be part of the needs.

3. Research Needs

a. Identify and implement cost-effective, presently available technology for reducing consumptive agricultural (in western states primarily), industrial, and municipal uses including power generation.
b. Identify, implement, and test prototype institutional arrangements needed to efficiently and comprehensively plan, develop, finance, and bring about efficient use of water resources on an area or watershed basis.
c. Develop cost-efficient technologies for developing new water supplies,

including groundwater. This includes the stabilization of water supply through water-short, water-rich, cyclic periods by use of recharge and other temporary storage technologies.

d. Develop new methods for on-site control of urban runoff.

e. Develop and adapt hydrologic-economics simulation capabilities on the multi-land-use large watersheds that can be used for: i) identifying water losses, major internal uses, and inefficiencies in water-use within the system, and ii) simulating the effects of alternative solutions for reducing water loss by improving water-use inefficiency or reallocating major water supplies within the system.

f. Develop new cost efficient technologies for reducing consumptive water uses in agriculture, industry and municipalities.

g. Develop technologies for providing water reuse up to and including closed loop systems, with special emphasis directed to the Year 2000 plus when additional water resources may be difficult or costly to obtain.

h. Develop "water farming" technology associated with water supply in reservoir and city areas. This would include evapotranspiration reduction on watersheds (e.g., grass for trees), water harvesting (asphalting) or water harvesting and storage (groundwater recharge from urban runoff).

B. Urban Water Quality

1. Current Situation

Urban water quality is affected by many factors. The complex array of land uses in an urban or urbanizing area generates a varity of pollutants. One broad scheme for categorizing pollutants of surface waters is based upon the kind of source. Two kinds of sources have been identified: point and nonpoint. Point sources of pollution are defined as concentrated discharges of waste water emanating from a specific, discrete site. Some obvious examples of major point sources include sewerage overflows, sewage-treatment-plant effluent, and industrial waste effluents. Nonpoint sources of pollution are defined as diffuse discharges of waste water that cannot be identified as coming from one discrete or point source. In this paper, emphasis will be placed primarily upon nonpoint sources of pollution that may adversely affect the quality of urban water supplies or runoff.

In many cases, nonpoint sources of pollution have assumed increasing relative importance as point sources of water pollution have come under increasing control in urban areas. In many urban areas, water-quality goals are not likely to be met unless nonpoint sources of water pollution are also controlled. Section-208 of The Water Resource Act amendments of 1972 planning activities and their implementation in urban areas has brought the need for sound scientific information on urban runoff into sharp focus. Implementation of some Section-208 plans has been hampered because the lack of precise information has made it difficult to define what is a water quality problem, what are the probably effective control methods and what are the probably results of putting a planned series of control measures into operation. A great deal of work remains to be done in the social- and the natural-science aspects of nonpoint pollution as it affects urban water quality.

2. *Statement of the Problem*

Pollution from urban runoff is a significant water-quality problem. Land-derived pollution sources are a very important component of that problem. The International Joint Commission notes that with nonpoint pollution, perhaps more than any other type, seemingly simple management decisions with respect to many diffuse sources may have complex ramifications that can include unintended consequences or even result in failure of the program concerned.

Specific urban water-quality problems have been well identified by several agencies. In addition, the EPA (1978) has identified five major categories of topic areas where knowledge is lacking regarding urban runoff. These are:

• Sources — Not enough is known about where pollutants originate. Major sources certainly include vehicles, vegetation, erosion, animals, and air pollution, but a better understanding of source contributions could enhance control opportunities.

• Washoff/transport mechanisms — Not enough is known about how pollutants get from the sources to the receiving waters. Models for simulating runoff would be more useful in problem definition and control evaluation if they could accurately reflect washoff and transport mechanisms.

• Impacts — It is still difficult to go beyond speculation in assigning urban runoff its proper share of responsibility for pollution problems where other sources also contribute. In cases where other sources create obvious problems, it is difficult to determine the appropriate level of control for urban runoff.

• Relative benefits — Local planners have difficulty deciding whether the various benefits of controlling urban runoff quality justify the costs involved. There is considerable controversy over the present dry-weather-standards' relationship to beneficial uses, given the time and space scales of storm events. Many plans fail to be implemented because of uncertainties regarding: How much control is enough? Who benefits? Who should pay? Who should decide?

• Controls — Some of the control measures cited in typical 208 plans are plausible candidates, but their application for the purpose of water-pollution control has not been studied quantitatively. Both cost and effectiveness data on full-scale programs are lacking. While the 5-

year work plan for nationwide urban runoff in operation by EPA aims to describe the nature of urban runoff problems, their cause, severity, and the opportunities and measures for controlling them, it is clear that the program is likely to identify many significant and specific problems that will require further scientific research as a basis for sound, practical actions by public and private agencies and organizations. From one perspective, we observe a problem that expresses and incorporates most of the EPA-enumerated needs for urban and nonurban areas.

Most land-management-program decisions that affect water quality are made by non-technicians who are concerned primarily with economic and socio-political impacts of these decision and may lack a good understanding of technical implications. However, for these decisions to be valid, they must be based on sound and credible technical knowledge. The problem is to provide technical credibility to decisions made by non-technical managers. The presently available models cannot be applied to real-world management decisions in this context because of model complexity, large and devise input data needs, and the lack of integrated economic simulation capability at the spatial scales where costs are incurred and political decisions are made. Presently, available hydrologic and chemical nonpoint source and transport models could be developed, simplified and combined with economic models to aid the capability for assisting decision making.

Many urban areas do not meet air- or runoff-water-quality standards because of wind borne dust. Much of this dust originates from wind erosion on agricultural lands. However, the effects of dust control could more likely have more import on urban than rural environmental quality.

3. *Research Needs*

a. Develop a decision assisting tool for water quality managers. Develop decision assisting tools from the water-quality- simulation models already available which can be used by water managers or design engineers for selecting among water-quality- control alternatives. Such tools must be accepted as valid from a political and economic standpoint. Also, this tool must be based on technically sound understanding of pollutant sources, transport mechanisms, and the effectiveness of control practices. An array of explicit supporting research needs which will be required for filling gaps in the needed understanding.

b. Evaluate the pollutant effect of fugitive dust in urban areas. Determine the import, sources, and, if warranted, effective control methods for controlling fugitive dust in urban areas. The complexity of the problem demonstrates the need to address environmental problems from a systems approach.

URBAN LAND

A. Land Use Change and Conflict

1. Current Situation

Every year, approximately 3 million acres of rural land are converted to urban or water uses. Of this total, about 600,000 acres are cropland. Additional rural land is idled by the urbanization process even though it is not actually built upon or directly used in an urban function. There are no good data on the amount of land that is idled in the urbanization process, but some informed experts have guessed that the amount idled may be double the amount actually converted. Conversion of rural land to urban uses has both economic and physical resource dimensions.

The character of urban expansion over time is tempered by many individual decisions on the use of privately held land. Changes in ownership and land use are made by individuals, each with different goals and aspirations as well as varying levels of stewardship and land ethic. The land product is also molded by land-use planners and by public investments such as highways, sewer and water systems, airports and others. Land-use changes, although rational from the view of the individual or public-agency decisionmaker, may impose significant costs on society that are not taken into account in the individual decisions. For example, highway construction may open up areas for development far from the edge of the city, and the resulting scattered pattern of development may impose high costs for provision of public services, eliminate the possibility of open-space or recreational uses of land close to the city, and generate land-use conflicts with the farm operations that are mixed with non-farm land uses. Or the decision of an individual or firm to construct a major shopping center on the urban fringe may have important effects on downtown business or central-city revitalization efforts may impose additional public service costs rural governments adjacent to the new development, may adversely affect nearby farm operations, and may encourage a more scattered development pattern on the urban fringe.

The conversion of rural land to nonagricultural use affects far more land than is actually built upon. Landowners, particularly farmers, who anticipate development in the future may change their patterns of farming, investments, and soil-conserving practices. Uncertainty about the future of agriculture in the area may lead to widespread disinvestment in farming; this has been termed the impermanence syndrome in agriculture. The impermanence syndrome may lead to special water-quality and soil-conservation problems near urban areas. Because their planning horizon may be short, landowners may have little incentive to invest in soil- and water-conservation practices to maintain soil quality and reduce runoff. Soils near urban areas may be mined , in effect, in anticipation of conversion of nonagricultural use. Large areas around major cities are affected, and if the population growth in rural areas between 1970-80 continues in the next decade, the impermanence syndrome may affect large areas of land near smaller cities in more rural areas.

Conversion of rural land to urban use also has important physical resource implications. In previous decades, the better lands were commonly used to support urban expansion. Steep lands, poorly drained lands, and lands subject to instability were often by-passed as development leap-frogged over these poorer quality lands and encroached onto agricultural land. However, as utility lines were extended to outlying developments and as larger earth-moving equipment was developed, the value and accessibility of previously by-passed lands increased. Many of these lands with soil and water constraints were brought under development. When development occurs without accommodating the soil's inherent limitations for development, the results may include wet or flooded basements, dying or oxygen-starved landscape plants, failing septic systems, polluted aquifers, unstable and erosion-prone yards, earth slides, foundation failures, structural damage by expansive clays, corrosion of concrete and metallic materials, and extremely acid or alkaline soil conditions. Local building and other codes have been developed over many decades of experience to accommodate the deficiencies in soils and lands for many types of land use. Except for multi-storied buildings, heavy structures, and facilities requiring engineering soil evaluations, most nonagricultural land-use decisions are made without any on-site investigation. The building codes, health regulations, and other local and state statutes are designed to provide some standard of performance for most uses. But the review and revision of these standards often lags behind the need to do so. Local officials often find themselves under the "tyranny of the imminent" – problems develop before they are anticipated, and codes are changed only after a need (crisis) is demonstrated.

2. *Statement of the Problem*

Several problems result from the resource — economic and physical — impacts of conversion of rural land to urban use. First, the extent and sources of land-use conflicts, and the costs to different groups in the society must be identified. For example, far more land is affected by the impermanence syndrome than will ever conceivably be developed; thus, the value of the land for agricultural purposes may be seriously reduced by erosion even though no development use is made of the property.

Second, water quality in the urban areas may be seriously affected; nonpoint pollution from farms in urban areas may be substantially greater than from farms in more rural areas. The magnitude of these problems is not well understood.

Third, the population must be educated, and scientists play a major role in this education effort. Articulate spokesmen are required to bridge the gap between the technician, scientist, and layman. In paraphrasing Glenn Frank, Clay Schoenfeld (1968) stated "The future of America is in the hands of two people — the investigator and the interpreter. We have a growing supply of investigators. But there is a shortage of readable and responsible interpreters, people who can effectively play mediator between specialist and layman." The education of the populace at all levels can perhaps be as effective in correcting many of our urban land problems as technological solutions.

Fourth, elected officials are faced with difficult policy choices on land-use controls, sanitary codes, the use of soils information in development permit decisions, and other choices without the research base to understand the consequences of the various policy choices, or easily applied methodologies to estimate magnitude and distribution of the benefits and costs in specific situations.

3. Research Needs

Several critical research needs emerge from the situation and problem analysis.

a. The process of conversion of land from rural to urban uses must be carefully studied to identify the major actors and the incentives to which they respond.

b. Economic and other social-science research is needed to identify the nature and extent of the impermanence syndrome. Closely related is the question of whether special soil conservation practices or programs should be developed for application to land owned by individuals who anticipate (incorrectly) the conversion of the land to nonagricultural use within, say 5 to 10 years. The long-run effect of lack of investment in practices to conserve soil quantity and quality on land near urban areas should also be investigated.

c. Research is needed to develop methodologies that can be used at the county or municipal level to identify the magnitude and distribution of the costs and benefits of alternative land use policies, including methodologies to estimate the costs of not attempting to influence the conversion of rural land to urban uses.

d. There is a need to develop programs, education curricula, and materials that provide land-use decisionmakers with the motivation to consider the capability of the land resource to accommodate proposed uses and determine environmental impacts of such uses.

B. Soil Characterization in Relation to Urban Constraints and Problems

1. Current Situation

A common attribute of soils in urban environments is that they are often disturbed; either through compaction, truncation, chemical alteration, or being "rebuilt". Urban "soils" are often derived entirely from materials transported by man to the site and comprised, in part, of artifacts of contemporary and past civilizations. The current soil taxonomy (Soil Survey Staff, 1975) recognizes these human-influenced soils in the Plaggepts and Arent suborders. However, recent experience and studies (Smith, 1976; Stein, 1978; Stein et al., 1977; Fanning et al., 1978) have shown the need for new and more refined diagnostic criteria and features to properly classify these soils for interpretations of their behavior.

The primary purpose of the soil classification program in the U.S. Department of Agriculture has been for use in making soil surveys (Kellogg, 1960). In addition to the soil-survey function of serving as a base for the application of technology in agriculture, soil surveys and other soil data have increasingly been applied to nonagricultural uses of soils (Bartelli et al., 1966; Buol et al., 1973; McHarg, 1969; Simonson, 1974).

The characterization and classification of urban soils in the context of current soil taxonomy is confined to the upper four to seven feet of the soil. This situation results in deeper substrata remaining outside or below the traditional discipline boundaries of pedology. It is within this zone of about seven to 25 foot depth that many of the constraints to and problems of urban land use occur (Miller, 1978; Robinson and Spieker, 1978). There is a great demand in urban environments for characterization of soil and substrata to depths below six feet for the purposes of planning land use and anticipating potential problems (Nichols and Campbell, 1971; Griggs and Gilchrist, 1977; Keller, 1979).

While geologists and engineers have often characterized these deeper materials and mapped their occurrence, these near-surface substrata maps are not available. Furthermore, their scale is usually not as detailed as planners and other land-use decisionmakers desire. Likewise, map-unit composition and descriptions are not refined enough for local or detailed planning. The confidence limits of data sets from each discipline vary enormously—often unknown to the user. As a result of this situation, many users of geotechnical information are interested in some kind of marriage between the disciplines of soil science (pedology), geology, hydrology, and other earth sciences so that nonagricultural land users can identify, anticipate, plan, and design for potential constraints and problems.

2. *Statement of the Problem*

There is a problem of defining soil — to the pedologist, it is that part of the earth's crust or regolith altered by biochemical processes, often corresponding to the rooting depth of the native perennial plants, where the processes of horizon development commonly terminate. To the geologist, soil is that portion of the weathered regolith, which may extend well beyond the pedologist's lower soil boundary. And to the engineer, soil is often considered as simply unconsolidated material that may extend well beyond the soil depth considered by both pedologists and geologists.

These different definitions and the taxonomies and characterization procedures that have developed within their respective disciplines have resulted in either overlapping or noncompatible data sets. Users of such information are often not aware of these differences. As a result, potential land-use problems are often overlooked and not anticipated.

Geotechnical data and information, such as soil surveys and geologic maps, are being used more intensively in areas pressured by nonagricultural land-use decisions. Differences in definitions, coupled with limitations and differences in accessibility, depth, scale, and confidence limits of these information resources often result in their being overlooked, misused, or used beyond their confidence limits. Furthermore, this situation can lead to contradictions in terminology and disagreements in interpretations, resulting in competition and mistrust among disciplines.

Solutions to these problems include interdisciplinary education and cooperation, the training of the users of earth-science information, clear statements or graphic displays of the depth limits and confidence limits of the information, and more intense characterization of the soil and substrata involved.

There are cases where unanticipated soil and geology-related land-use problems are encountered, either because the problem was beyond the scale or confidence limits of the geotechnical data sets, or because the map-unit composition was not well defined or characterized. It is not uncommon in such cases that the conditions contributing to the problem were known to earth scientists, but because information formats, publishing policies or characterization techniques were inadequate to accommodate such knowledge, "it fell through the cracks." Without greater attention to such concerns, the future is not likely to be much different.

3. *Research Needs*

Research is needed to improve taxonomies and characterization data in all earth-science disciplines to preclude the aforementioned problems. Perhaps the greatest contribution can be in the mode by which the data and information are presented and displayed. Research needs break out into three categories:

a. Data acquisition, soil-substrata characterization, and mapping techniques to determine the map-unit composition, variability, and confidence limits of existing maps and data sets.
b. Interdisciplinary techniques to display soil and substrata materials to appropriate depths, and packages of these and other data in useable formats to aid in land-use policy decisions.

c. Education-training of land-use decisionmakers to utilize earth-science information and institutionalize such information (e.g., through building, health codes, etc.).

C. Soil as a Medium for Plant Growth

1. Current Situation

Social attitudes in urban areas still reflect the importance of herbaceous plants, shrubs, and trees in the urban environment (Patterson, 1974; Schmid, 1979). The design, planting, and maintenance of urban vegetation are a multi-billion-dollar industry. Considerable funding for the establishment and maintenance of this vegetation is derived from public tax monies to support public parks and gardens, recreation areas, street plantings, roadside vegetation, public building grounds, etc.

Most plants used in urban areas were not selected for their tolerance to the often-hostile urban environment. Rather, they have been selected because of their ease of propagation. As a result, establishment and maintenance of urban plants and resultant high mortality rates contribute to high costs of plant cultivation.

Schmid (1979), Duell (1979), Santamour (1976), and others have reviewed the types of urban and transportation right-of-way vegetation, their role, and the constraints affecting such vegetation. Often, an unfavorable soil medium in urban areas and along road right-of-way is one of the major constraints to good vegetation performance. These soil-related problems include compaction; poor aeration due to fine texture, high density, or water saturation; moisture stress; nutrient imbalance; temperature extremes; conditions conducive to insects and pathogenic organisms; toxic quantities of one or more compounds; and the synergistic effects of two or more of these. Often, little or no evaluation of existing conditions is made prior to planting.

Toxic conditions include those caused by exposure of unoxidized sulfide-bearing soil materials, residues from demolition projects and burns (e.g. lead from paint), dredge spoils, deicing salts, fuel spills, and other toxic compounds within or added to urban and man-influenced soils, including air-pollution residues. Often the circumscribed rooting space allocated to urban trees and other plants, along with other poor soil conditions, subjects plants to physiological stress. Such stress magnifies the impacts of diseases, insects, air pollution, and limited water supply, which otherwise could be tolerated by many species. The narrow genetic lineage and relatively few species of plants relied upon by managers and planners of both public and private urban vegetation defies a basic tenet of biological stability — diversification. Dutch elm disease and chestnut blight provide stark lessons in reliance upon single species performance. And yet, the hostility of many urban plant environments is strong incentive for plant managers to rely on those species that have proven reliability in the often-harsh urban environment.

Many soil-related problems can be predicted. Prior knowledge of site conditions can be obtained through soil surveys (where available), geologic maps and data, previous studies and historical records, or on-site analysis. The information in soil surveys and geologic maps is not site specific and should not be used for design purposes. But such information can provide insights on the general constraints and potentials of the site for planning purposes. Too frequently, however, none of these sources is consulted. On-site analysis often occurs after a problem has already arisen. There is no substitute for on-site evaluation of soil conditions prior to design and planting. Retrofitting the soil environment to the plant requirements is beyond practicality and budgets in most instances.

Political jurisdictions, through their planning, public works, permits, or other departments, are requesting information and assistance on developing soil specifications for plant-growth media such as topsoil . Similar inquiries involve specific soil conditions that require alteration prior to use. For example, saline dredge spoils, acid mine spoils, and other sulfide-bearing materials have caused considerable anguish in many areas for both private and public land managers. These conditions are often recognized by soil scientists, geologists, plant scientists, and others and information about them is often recorded in various documents. But because this knowledge is not institutionalized within building codes, regulations, and ordinances, innocent misuse of these tainted soil materials occurs on the part of the developer, homeowner, public land manager and others. Millions of dollars in plant damage and loss could be saved annually by simply consulting available information and testing for site-specific conditions.

2. *Statement of the Problem*

Failure to address the soil and plant problems in urban areas will result in unnecessary expenditures of both public and private monies. The urban landscape will suffer aesthetically and environmentally from these losses. Those responsible for contracting public projects and plantings should be trained to demand site-assessment information and plant-tolerance criteria prior to the letting of contracts. Minimum site criteria (e.g. topsoil criteria, physical and chemical criteria for subsoil) should be developed to accommodate proper site assessment.

Soil amendments from urban waste products (e.g. sludges, composted wastes, leaf composts) are being used. These uses stem from two pressures: First, the need to dispose of or recycle urban-generated wastes, and second, to supply some nutrients and organic matter to the root zone in the face of increasing costs for commercial fertilizers. These waste products are not innocuous, however. Additional information is necessary regarding the long-range impacts of using these products and the tolerances of various plant species to these materials.

Correcting the aforementioned problems will save billions of dollars in the future through longer-lasting vegetation, reduced maintenance and replanting costs, and an enhanced urban environment.

3. Research Needs

Research is needed to evaluate materials and methods for modifying the soil and the effects of these modifications with time.

D. Soil as a Sediment Source

1. Current Situation

Sediment from urban areas is a major pollutant of surface waters that originate in or flow through such areas. The water quality of more than half of the nation's drainage basins is affected by urban runoff.

While sediment from urban sources is very important locally it constitutes only 4 percent of the estimated total sediment contribution attributed to all land uses in the United States (USDA, 1980).

Once sediment passes from the land into the water system, the results are restricted flow and reduced storage capabilities, possible cessation of biological activity, navigational restrictions, and ultimately the problem of clean-up and disposal.

Sediment sources, such as construction sites, are often temporary in nature and concentrated in relatively small portions of a metropolitan area.

As point-source pollutants from municipal and industrial sewerage systems are increasingly controlled, the nonpoint-source pollutants become increasingly important. Suspended soils or sediment constitute roughly 90 percent by weight of pollutants in urban storm-water runoff. In turn, sediment and other suspended solids in urban runoff currently make up 55 percent of the total weight of all pollutant loading, both storm water runoff and sewerage discharge for the urbanized United States (USDA, 1980). This is expected to increase to 66 percent by 1990 due to expected decreases in the weight of pollutants derived from combined sewers and secondary and advanced sewage-treatment facilities (USDA, 1980).

Sediment, while great in volume and in weight, may not cause water pollution in proportion to its concentration or volume when compared to other pollutants such as toxic metal and organic compounds, rapidly oxidizable organic matter, and nutrients such as nitrogen or phosphorus. The turbidity caused by sediment and the infilling of channels and reservoirs can be very serious problems, however.

2. *Statement of the Problem*

Some sources of sediment and their transport mechanisms are not clearly understood, and cost-effective, on-site means for controlling the sources, once identified, are as yet imperfect. Sediment from construction sites is identifiable and relatively controllable by changes in site-management practices. Yet many local governmental units and private land developers have failed to implement programs and practices needed to reduce erosion in order to control the amount of sediment entering streams and lakes in urban areas.

3. *Research Needs*

a. Design models that will allow predictions of delivery ratios of sediment from urban erosion sites given contracting sites and situations.
b. Evaluations of the cost-effectiveness of alternative on-site erosion control practices both existing and newly-developed.
c. Determine technical, social, institutional, financial, and other barriers that prevent greater adoption of erosion- and sedimentation-reducing techniques by public officials and private land developers.

E. Soil as a Medium for Waste Recycling and Renovation

1. *Current Situation*

There is a growing volume of both liquid and solid waste materials generated by society that are being applied to or placed in or under the soil. Disposal of these materials on the land has grown rapidly during the decade of the 1970's and shows every indication of continuing to grow during the 1980's and beyond. Awareness of the toxic components in many of these wastes and their potential for environmental damage and health impacts when improperly handled has increased notably in recent years. Because of transportation cost, most liquid and solid wastes for purposes of disposal or attenuation, are placed on or in the land close to urban areas.

As the use of land for this purpose increases, there is a strong need for thorough understanding by many individuals and organizations of the principles and processes involved when wastes interact with the various components of the soil.

During the past decade there has been a striking increase in research on many of the physical, biological and chemical aspects of disposal of liquid wastes in and on the soil (Volk and Landa, 1979). Much research has dealt with specific, discrete processes or subprocesses of a physical or biological nature, while other research and demonstrations have dealt with broad systems approaches to waste disposal and renovation. Sopper

(1979) summarized the current situation in the United States regarding surface application of sewage effluent and sludge. He gave special attention to the substantial experience gained through the Pennsylvania State University Comprehensive Demonstration Program. The importance of a comprehensive approach to planning waste disposal and renovation, including a planned system for water-quality monitoring, was stressed by Sopper. He also noted that land application of municipal waste water and sludge can result in beneficial side effects due to the conversion of waste material into materials that will enhance the growth and development of economic crops, the purification of toxic and hazardous liquids, and the recharging of ground-water systems.

The on-site, subsurface application of sewage effluent from households and other small sources is a second major use of soil as a medium for waste attenuation. Such systems are widely used throughout the United States in households beyond the service area of the central sewerage collection and treatment systems found in metropolitan areas.

The 1975 EPA Report to Congress (USEPA, 1975) stated that almost 20 million housing units, representing about 29 percent of the United States population, dispose of domestic wastes through individual on-site disposal systems. About 85 percent of these systems are septic tanks and cesspools, which produce approximately 800 billion gallons of waste per year.

A wide variety of environmental and health problems have developed during the past three decades as inadequately planned and designed systems were used in households, farms and small businesses that generated these wastes. Often, the soil medium and system design are inadequate to accommodate the volume and strength of these wastes using current disposal systems. Bouma (1979), has described the basic principles, the kinds of alternative systems available, and the use of soil-survey information to extrapolate experimental findings related to these kinds of sewage disposal and renovation systems. He summarizes the results of more than a decade of intensive research and demonstrations on the physical, chemical, and biological aspects of these systems.

2. *Statement of the Problem*

Some of the problems relating to soil as a medium for waste disposal or attenuation are as follows:

a. Our understanding of physical, chemical, and biological interactions of wastes with soil systems is inadequate.
b. There is a need for growth in understanding and use of planning models among state and local officials who must deal with the problems of a planning process for waste disposal in and on soils.
c. The phenomenal growth in the volume and diversity of waste products produced by our society has created fears and uneasiness among many

in the general public about what disposal systems, if any, can deal safely with these waste materials on a long-term basis. There is a major need for better public understanding of desirable alternative ways of handling wastes.

d. The institutional mechanisms at the federal, state and local levels for dealing with waste materials are improving but are still inadequate for making timely and effective decisions.

3. Research Needs

a. Research is needed in the physical-sciences area to better relate information from soil mapping and classification systems to the planning and design of subsurface sewage-disposal systems on a variety of specific sites. In addition, there continues to be a strong need for research in the institutional, legal, and economic areas regarding optimum use of these systems.

b. Modeling of the long-term performance of selected sytems in defined settings will be particularly useful in predicting with confidence the long-term effects upon ground water resources of waste disposal in and on the soil.

c. Health aspects of such systems need further research, especially pathogen transport and attenuation in conventional and novel systems in soils with large structural voids.

d. Research on engineering design to achieve water conservation through flow reduction and waste-water segregation is needed.

F. Other Soil-Landscape Constraints

1. Current Situation

Billions of dollars and untold human frustration and trauma result from soil failures due to uses of the soil that are not commensurate with the soil's capability to accomodate the uses or loads. This situation will continue to occur without increased attention to site evaluation and prediction of soil behavior. Much of this damage is avoidable now if current geotechnical information is needed.

Engineering uses of soil materials and their behavior in response to loads properly belongs to the disciplines of soil engineering and soil mechanics. But other disciplines such as soil science, geology, chemistry and hydrology also have an interest in engineering uses of soil and have contributed to the knowledge of soil behavior under engineering usage (Simonson, 1974; Legget, 1967; Nichols and Campbell, 1971; Robinson and Spieker, 1978).

Since most nonagricultural uses of land are initiated without any on-site investigation, earth scientists are commonly asked to predict soil behavior under low-load structures are often placed within the soil profile defined by pedologists. Soil scientists, in describing, classifying, and characterizing the pedogenic soil profile, make engineering predictions of soil behavior in conjunction with soil engineers. Engineering design criteria, however, are the domain of the engineer.

For most low-load soil uses, where on-site analysis is cost-prohibitive, local building codes are relied upon to provide some degree of insurance against soil failure. These codes, specifying such items as depth, width, and type of footing and sub-base material, have been formulated by engineers and people in other disciplines based on both local experience and engineering principles.

In order to protect the public safety, building codes establish minimum design criteria to insure some degree of structural stability and to reduce the risk of failure. As urban areas expand, the probability of encountering problem soils and geologic conditions increases. Building codes, therefore, must be periodically reviewed and revised. Furthermore, soil conditions involving chemical, mineralogical, and microbial reactions are encountered which may not be accounted for in routine engineering evaluations. Thus, building codes must accommodate both engineering input and input from various earth-science disciplines. The expansion of urban land uses and associated capital improvements (e.g. rapid-transit systems, cross-county water and sewer mains) occasionally encounter soil conditions where the constraint is a chemical or mineralogical condition rather than a limitation of soil stability or structural bearing capacity. Engineering design of the structures may be sound, based on standard testing procedures. But if the earth material contains sulfide minerals, and oxidation is allowed to occur through exposure or by artificial drainage, serious acid-producing reactions can take place with subsequent structural damage. Concrete and metallic components placed within these sulfidic materials are subject to severe corrosion and dissolution.

Since chemical and mineralogical evaluations of earth materials are not routinely done, these conditions can be and, at times, are overlooked. Housing developments and other projects have been installed in or on these acid-producing sulfidic soils. Excavations into such material expose fresh sulfides that begin rapid oxidation and sulfuric acid production, which results in very low pH values and toxic and corrosive conditions for foundations and pipework. proposals have been made to ameliorate some of these acid soil conditions with applications of municipal sludge for the purpose of establishing vegetation. However, the acid condition coupled with acid-mobilized metals within the sludge may compound the problem by generating potential health and environmental problems.

Political jurisdictions faced with such crises request the assistance of soil scientists, geologists, engineers, and others in not only mitigating an existing situation, but in revising current codes to identify the location of such materials and promulgate regulations for using such sites. Often geologists, soil scientists, and others are aware of potential subterrainean strata or horizons that can cause structural, chemical, and mineralogical problems, but because of map scale, map-unit-composition limitations, discipline boundaries, and publication policy and formats, these conditions are not always identified on geologic maps or described in soil surveys. A crisis is commonly necessary to prompt local governments to convene a panel of earth scientists to recommend courses of action to formulate stricter land-use regulations that will identify such conditions and require design criteria to accomodate the condition.

2. *Statement of the Problem*

Part II of Resources Conservation Act draft report cites the study of Alfors et al. (1973) which indicates that between 1970 and 2000, losses from landslides in California alone will cause nearly 10 billion dollars in property damage unless present soil management practices change. Soil-induced problems such as cracked foundations; ruptured and damaged utility installations; road failures; dam, dike and embankment failures; corrosion and dissolution-induced structural stresses; and other structural damage cost billions of dollars annually in the United States.

Wet soil conditions and water management are problems in many urbanizing areas. About 27 million acres — more than one third of all urban areas— are developed on wet soils and substrata. Development on these lands is conducive to wet basements, malfunctioning septic systems, poor plant-growth conditions, structural damage from frost action, and high maintenance costs for driveways and streets (USDA, 1980).

Corrosivity is another soil problem requiring additional research. Resistivity is the commonly used measure by which the corrosivity of earth materials is determined. Despite the common usage of resistivity as a parameter to predict soil corrosion potentials, there are many experiences where resistivity is not a reliable index of potential soil corrosivity (van Eck, 1966).

Corrosion in a particular soil is often attributable to several soil properties that interact to produce the soil's corrosivity aggressiveness. Much more research and data are necessary to increase the reliability and confidence limits of soil corrosivity predictions. Earth scientists need to make more measurements of in situ resistivity on landscapes and correlate corrosivity with other soil attributes that will enhance the confidence limits of soil survey and geologic interpretations.

There is a need for interdisciplinary investigations and information display in this field. Land-use decisionmakers are usually faced with specific decisions and do not recognize discipline boundaries. The result is that data are often misused, misapplied, extrapolated beyond the discipline boundary, or pushed beyond their confidence limits.

Soil scientists, geologists, engineers and other earth scientists need to examine the potential for providing structural support information and identifying unstable and corrosive environments. Sampling frequencies, quantified map unit composition information, and data on the confidence limits of such information are necessary to establish the potential for providing such behavior predictions. It may be found that the variability of the soil and substrata preclude such attempts at structural behavior interpretations. If so, the user of soil and geologic maps and other geotechnical information should be clearly warned of the potential constraints and the inability to predict structural load behavior with any degree of precision.

Whether earth scientists are able or not to provide the land-use decisionmaker with quantitative information on soil and substrata structural behavior, at least qualitative assessments will provide the decisionmaker with useful information. Land-use constraints will be identified or their potential existence pointed out. The land user then will be provided with enough information to determine whether further investigation is necessary.

3. Research Needs

a. Interdisciplinary research is needed to characterize, develop confidence limits, and display soil and substrata materials to appropriate depths and to package this and other necessary information into useable formats and develop methodologies to aid in land-use policy decisions.

b. Research is needed to develop techniques, strategies, and methods to correct or modify problem-prone soils so that urban uses can be made of these lands without structural failures and damage or environmental damage.

RECREATION

A. Urban Recreation

1. Current Situation

Recreation including activities engaged in during non-work time is a land and water use ranging from the urban community to rural, remote, undeveloped areas (wilderness). In the last 35 years demand for recreational uses of land and waters has greatly increased. This situation is enhanced by increased income, leisure time, mobility, and better access to opportunities. The public sector, particularly, has provided developments in city lands and waters for the public — i.e., parks, forests, trails systems, rivers, lakes and streams.

With increasing costs and declining amounts of fossil fuels recreation may well be more concentrated in the urban and new urban location—

where now over 70 percent of our population resides. Thus, it can be anticipated that recreation and recreationists will compete heavily for use of land and water with other entities in the urban community. Such recreation activities include such land uses as: camping, picnicking, hiking, photography, skiing, snowmobiling, and sports; and water uses are: boating, canoeing, fishing and sailing.

2. Statement of the Problem

From the standpoint of soil and water resources, urban recreation developments may present erosion and water-quality problems depending on soil depth, type, and characteristics; water supply (quantity and quality); sewage disposal; vegetation; topography; and shoreline (of lakes or streams). Proper maintenance of area soil and water resources as well as maintaining the quality of the recreation experience requires monitoring and evaluation of life impact. This includes water quality studies as well as the determination of the effect of use on soil and vegetation. Study results should be presented in ways identifiable and applicable by resource managers.

3. Research Needs

Study of the environmental impacts of recreational uses of urban land and water resources to help resource managers and planners determine site degradation and avoid overuse of recreation resources.

B. Wildland Recreation

1. Current Situation

Moving from urban areas on a continuum extending to the most remote, undeveloped land, appearing first are county recreation areas, and state parks. Many of these have problems similar to those of the city. Here, soils may be too shallow for development or highly erodible. Today, many county and state parks contain eroded areas due to overuse.

One of the activities in wildland recreation that spans the whole gamut from the nearer urban areas to the undeveloped wilderness, except those lands designated under the National Wilderness Preservation Systems, is the use of various off-road recreation vehicles (ORVs). These are used on lands and some parks throughout the country, not only on roads and trails, but also cross-country. There have been some studies on impacts such as those dealing with the effect on the southeastern Montana rangeland (Leininger and Payne, undated ca. 1980). There are particular problems with ORV impacts on vegetation, soils and watersheds in the desert areas in the southwest and California (Bury et al., 1976).

On recreation lands there are concerns as to adequate quantity and quality of water for drinking, for bathing, and for washing and general use. Other concerns include siting for sanitation systems on or near lakes and streams to various kinds of pollution, both within the water body and from nonpoint-pollution sources (Garn and Parrott, 1977).

Specific to designated sites within these remote areas are problems of impacts on development sites such as camping areas. Other concerns are heavy use, trampling, nonpoint water-pollution sources, toilet location, and shallow soils that will sustain only limited development.

There is a whole series of questions related to dispersed recreation on wildlands, including remote trailed areas and wilderness within the Wilderness Preservation Systems. A 1978 conference on "Recreational Impact on Wildlands", held in Seattle, Washington, produced a proceedings of 57 papers concerned with the perceptions of impact, research findings and experiences in preventing and remedying impacts plus approaches to future area management (Ittner, 1978). With respect to soil, the key things influencing wilderness qualtiy for recreational use are trafficability, soil depth, drainage and erodibility. Erodibility is a concern on trails and around lake campsites, particularly in high mountain areas.

Dispersed outdoor recreation in remote locations is appealing to many people. However, where there are limited facilities the potential exists for unfavorable impacts on area soil and water resources because of possible introduction of fecal coliform into streams and lakes near campsites.

In Rocky Mountain National Park, Colorado, there are problems with outbreaks of giardia. Total coliform in some streams has increased to the point of human hazard with management recommending boiling or treatment of all water used for human consumption (Ittner, 1978, p. 329). It appears that in some areas weather conditions are a major factor in moving pollutants into streams.

Some recreation areas in eastern Washington were found to be affected by the presence of cattle grazing nearby. This apparently increases the number of fecal bacteria in surface water. Rain accelerates the movement of fecal bacteria from soil to surface water. Study in Washington's North Cascades in 1973 showed that indicator bacteria tend to survive longer in cold water emphasizing the delicate nature of high-quality waters and the need for strict sanitary guidelines to insure their future safety and quality.

2. Statement of the Problem

Recreation users impair soil and water resources, which in turn influence recreationists. These relationships need to be investigated in specific situations by means of particular studies. There are several major studies suggested in the soil and water area:

a. Basic studies of recreationist impacts on specific sites—parks, streams, lakes — as they affect soil and water should continue.
b. A major concern throughout the United States, particularly in the West, is the use and impact of outdoor recreation vehicles (ORVs). For the West it seems important to study their impact on soils, particularly in the desert areas and where there is very limited moisture and where impacts will often last for many years.
c. Another important area of study is water-use and water-quality in lakes and streams of remote areas.
d. A fourth concern is nonpoint-pollution impacts caused by pollutants from recreational lands as well as from farm and rangelands being washed into the water during storms.

This is where recreation comes up against agriculture. In the Upper Midwest, where lakes are very valuable for recreation purposes such as fishing, and boating, many are being affected by nearby agriculture. Here it is very important that the microbiologists, the soil scientists, the hydrologist, and the recreation planner work to understand some of the major problems in dealing with soil and water resources for recreation purposes. Scientists can cooperate to obtain answers that will be useful for solving problems related to recreation impacts, both within the urban sector, the non-urban lands, and remote wildlands.

3. Research Needs

a. Studies of the impact by recreationists on soil and water resources used for dispersed and developed recreation (parks, lakes, etc.) particularly by ORVs to determine carrying capacities for such resources.
b. Studies of water use and quality in remove recreation areas including the effects of nonpoint pollution.

RESEARCH PRIORITIES

Most Americans live in urban areas. One highly visible land-resource issue is the conversion of rural land to urban uses. Land-use changes, although rational from the individual's or agency's views, may degrade the quality of soil and water resources, ignore the physical constraints of the land, and impose economic and social costs on the community, state, and nation. Thus, the rights of owners may conflict with the rights of the community.

Converting agricultural land to urban use eliminates the option of using this land for future agricultural production. Conversion increases pressures to intensify production on the remaining acreage, which has important implications for soil conservation.

Urban activities affect water quantity and quality. As urbanization proceeds, flooding becomes more severe and growingdemands for urban water compete with those of agriculture and recreation. To the extent that urban water-use efficiency is increased, more water is available for other uses.

Urban areas generate large amounts of a variety of wastes that can pollute surface and ground waters. One of the major sources of surface-water pollution is municipal sewage effluent. As improved treatment of sewage occurs, non-point pollutants from urban areas, including sediment, assume far greater relative importance.

Urban wastes are often applied to soils and may have value as sources of nutrients and organic material. But land application may contaminate ground and surface waters with pathogens, heavy metals, and other toxicants. We need more information on the interaction of some waste components with the soil-plant-water system to design safe, effective land disposal systems.

Large portions of urban areas are built on problem-prone soils, including some that are unstable, wet, or corrosive. Damage to buildings and other problems resulting from these conditions cost billions of dollars annually.

Soils in urban areas may impose severe stress upon vegetation. Urban vegetation is pleasing, may produce food, and insulates against cold and heat, and absorbs noise. A favorable rooting environment is often lacking and a favorable rooting medium must then be created. Innovative soil-amendment techniques, including the use of urban-generated wastes, are needed. Our present knowledge of the wide range of microenvironments in urban areas is inadequate.

Many of the physical problems of urban development result from individual and group decisions based on inadequate knowledge or assessment of the land's capacity to accommodate various uses. Much of this knowledge gap results from a mind-set that views land merely as space upon which proposed uses are located. Science-based education can sometimes be as effective as technological solutions in preventing misjudgments about the use of urban land. Demand for recreational land and water resources has increased strongly in recent years in both remote and urban areas. Activities such as fishing, boating, camping, hiking, hunting, and snowmobiling have a direct impact on the environment. Where the intensity of the activities exceeds the capacity of the land or water resource to tolerate these uses, environmental damage results. Research needs identified were:

1. Increase Water-use Efficiency and Water Supply

There is an urgent need to apply known technologies and develop new ones for increasing water-use efficiency and supply. This research needs to be applied to:

a. Reducing consumptive uses of water for urban, agricultural, and energy-related purposes.
b. Developing and testing alternative institutional arrangements at the political jurisdiction or watershed level to make more efficient use of water.
c. Developing cost-efficient technologies for augmenting water supplies, including groundwater. This includes the stabilization of water supply through water-short and water-rich cycles by use of recharge and temporary storage.

2. Simplify Available Nonpoint-source-pollution Models for Decision making Purposes

These models are needed by elected officials and others to decide among water-use alternatives on sub-watersheds, watersheds and large river basins. The research needs include:

a. Further developing source, transport, and control-practice components of existing urban and non-urban nonpoint-source models that provide an understanding of the hydrochemical systems involved but are too complicated to use as decisionmaking tools;
b. Developing simplified models by identifying critical components and making simplifying assumptions so that less input data are required and computation is easier. This would make the models usable by real-world planners,
c. Evaluating the simplified models in urban settings.

3. Analyze Urban Land Conversions and Related Land Use

Research on the processes of urban-land conversion is needed to determine the economic, social, and energy costs and benefits of alternative policies, including maintaining the status quo.

4. Delineate Problems of Urban Soils and Define Solutions for Land Uses Involving Construction or Waste Disposal

Interdisciplinary research in soil science, geology, and hydrology is needed for:

a. Characterizing the ranges in properties and spatial variabilities of soils and underlying geological materials and expressing them in terms of probabilities or confidence limits.
b. Developing techniques for modifying land uses or construction methods to compensate for inherent soil problems.
c. Developing models for predicting long-term and large-scale effects of water-disposal systems on soil, surface water, and ground water.

5. Evaluate Relationships Between Recreational Land Uses and Environmental Impacts

The research need is to measure the environmental impacts of intensive forms of recreational activity on land and water resources and to develop systems (including predictive models) that will help resource planners determine the susceptibility of a given site to degradation.

REFERENCES

1. Alfors, J.T., J.L. Burnett, and T.E. Gay, Jr. 1971. Urban geology-master plan for California. Calif. Div. of Mines and Geol. Bull. 198.
2. Barlowe, R. 1975. Demands on agricultural and forestry lands to service complementary uses. p. 105-121. Perspectives on prime lands. USDA, Washington, D.C.
3. Bartelli, L.J., A.A. Klingebiel, J.V. Baird, and M.R. Heddleson (eds.) 1966. Soil surveys and land use. Soil Sci. Soc. Am. and Am. Soc. of Agronomy, Madison, Wis.
4. Buol, S.W., F.D. Hole, and R.J. McCracken. 1973. Soil genesis and classification. The Iowa State University Press, Ames.
5. Bury, R.L., R.C. Wendling, and S.F. McCool. 1976. Off-road recreation vehicles—a research summary. 1968-75. Texas Agric. Exp. Stn. MP-1277.
6. Bouma, J. 1979. Subsurface applications of sewage effluent. In M. T. Beatty, G.W. Peterson, and L.D. Swindale (eds.) Planning the uses and management of land. Agronomy 21:665-703. Am. Soc. of Agron., Madison, Wis.
7. Dideriksen, R.I., A.R. Hidlebaugh, and K.O. Schmude. 1977. Potential cropland study Statistical Bulletin No. 578, Soil Conservation Service, USDA.
8. Dubos, R. 1976. Symbiosis between the earth and humankind. Science 193:459-462.
9. Duell, R.W. 1979. Vegetation for transportation systems and utility corridors. In M.T. Beatty, G.W. Petersen, and L.D. Swindale (eds.) Planning the uses and management of land. Agronomy 21:581-607. Am. Soc. of Agron., Madison, Wis.
10. Fanning, D.S., C.E. Stein, and J.C. Patterson. 1978. Theories of genesis and classification of highly man-influenced soils. Paper presented before International Soil Science Society, Edmonton, Alberta, Canada (Manuscript available from Dep. of Agronomy, Univ. of Maryland, College Park).
11. Garn, H.S., and H.A. Parrott. 1977. Recommended methods for classifying lake condition, determining lake sensitivity and predicting lake impacts. U.S. Forest Service, Eastern Region, Hydrology Paper 2.
12. Griggs, B.B., and J.A. Gilchrist. 1977. The earth and land use planning. Duxbury Press, North Scituate, Mass.
13. International Joint Commission. 1980. Pollution in the Great Lakes Basin from land use activities. An IJC Report to the Governments of the United States and Canada.
14. Ittner, Ruth (ed.). 1978. Water quality. pp. 329-331. In Proceedings, Conference on Recreational Impact on Wildlands. U.S. Forest Service No. R-6-001-1979.
15. Keller, E. A. 1979. Environmental geology. 2nd ed. Merrill Pub. Co., Columbus, Ohio.

16. Kellogg, C. E. 1960. Soil classification, a comprehensive system - 7th approximation. USDA, U.S. Government Printing Office, Washington, D.C.

17. Legget, R. F. 1967 . Geology and engineering. (2nd ed.) McGraw-Hill Co., New York.

18. Leininger, W. C., and G. F. Payne. undated (ca. 1980). The effects of off-road vehicle travel on rangeland in southeastern Montana. Montana Agric. Exp. Stn., Res. Report 153.

19. McHarg, I.L. 1969. Design with nature. Natural History Press, Garden City, N.J.

20. Miller, F. P. 1978. Soil survey under pressure: the Maryland experience. J. Soil Water Conserv. 33(3):104-111.

21. Nichols, D. R., and C. C. Campbell (eds.). 1971. Environmental planning and geology. U.S. Geological Survey and U.S. Department of Housing and Urban Development. U.S. Government Printing Office, Washington, D. C.

22. North Atlantic Regional Water Resources Study Coordinating Committee. 1972. North Atlantic Regional Water Resources Study. Appendix R, Water Supply. Prepared by North Atlantic Regional Water Resources Study Group, North Atlantic Division, Corps of Engineers, U.S. Army.

23. Patterson, J.C. 1974 Planting in urban soils. Ecological Services Bull., No. 1. U.S. Department of Interior, National Park Service. U.S. Government Printing Office, Washington, D.C.

24. Robinson, G. D., and A. M. Spieker (eds.). 1978. Nature to be commanded— U.S. Geological Survey Professional Paper 950. U.S Government Printing Office, Washington, D.C.

25. Santamour, F.S., Jr. (ed.). 1976. Better trees for metropolitan landscapes. U.S. Department of Agriculture Forest Service Gen. Technical Report NE-22. U.S. Government Printing Office, Washington, D.C.

26. Schmid, J.A. 1979. Vegetation types, functions, and constraints in metropolitan environments. M. T. Beatty, G. W. Petersen, L.D. Swindale (eds.). Planning the Uses and Management of Land. Agronomy 21:499-528. Am. Soc. of Agron., Madison, Wis.

27. Schoenfield, C. 1968. Educating the public in natural resources. J. Soil Water Conserv. 23 (6):224-227

28. Southeastern Wisconsin Regional Planning Commission. 1978. Planning Report 30. Regional Water Quality/Management Plans for Southeastern Wisconsin. Vol. 1, Inventory Findings. Southern Wisconsin Planning Commission, 916 N. E. Ave., Waukesha, Wis.

29. Simonson, R. W. 1974. Non-agricultural applications of soil surveys. Elsevier Scientific Publishing Co., N.Y.

30. Smith, H. 1976. Soil Survey of District of Columbia. USDA Soil Conservation Service in cooperation with USDI National Park Service. U.S. Government Printing Office, Washington, D.C.

31. Soil Survey Staff. 1975. Soil taxonomy. USDA Handbook No. 18. U.S. Government Printing Office, Washington, D.C.

32. Sopper, W. E. 1979. Surface application of sewer effluent and sludge. In: Beatty, M. T., G. W. Petersen, and L. D. Swindale (eds.). Planning the uses and management of land. Agronomy 21:633-663. Am. Soc. of Agron., Madison, Wis.

33. Stein, C. E. Mapping, classification, and characterization of highly non-influenced soils in the District of Columbia. M. S. Thesis, Department of Agronomy, University of Maryland, College Park.
34. D. S. Fanning, and J.C. Patterson. 1977. Variability and inclusions in mapping units of urban soils in river dredgings and miscellaneous fill. Am. Soc. of Agron. Abstr. p. 131.
35. Susquehanna River Basin Study Coordinating Committee. 1970. Susquehanna River Basin Study. Appendix F, Water Supply and Water Quality.
36. U.S. Department of Agriculture. 1980. Appraisal 1980. Review Draft, Part II. Soil and Water Resources Conservation Act U.S Government Printing Office, Washington D.C.
37. U.S. Environmental Protection Agency. 1975. Report to Congress. Waste Disposal Practices and Their Effects on Groundwater. U.S. Government Printing Office, Washington, D.C.
38. —. 1978. 1978-1983 Work Plan for The Nationwide Urban Runoff Program. Water Planning Division, EPA, Washington, D.C.
39. U.S. Department of Interior, Bureau of Outdoor Recreation. 1973. Outdoor Recreation: A legacy for America. U.S. Government Printing Office, Washington, D.C. p. 24.
40. van Eck, W.A. 1966. Soil science for corrosion engineers. p. 125-137. In Proceedings of the 11th Annual Underground Corrosion Short Course. West Virginia Univ. Engineering Exp. Stn. Bull. No. 80.
41. Volk, V. V., and E. A. Landa. 1979. Principles and processors involved in waste disposal and management. In M. T. Beatty, G. W. Petersen, and L. D. Swindale (eds.). Planning the uses and management of land. Agronomy 21:611-631. Am. Soc. of Agron., Madison, Wis.

9

William A. Berg and Charles M. Smith,
Co-Chairmen

Earl F. Aldon, Iris Ballew, Robert L. Beck,
Richard L. Brammer, Willie R. Curtis,
H. P. "Trey" Harbert III, Sigurd W. Melsted,
Robert J. Willard

DRASTICALLY DISTURBED LANDS

Conservation Needs, Technology, and Policy Alternatives

Drastic land disturbance results in intense soil, water, land use, and aesthetic alterations. These, in turn, affect a much larger off-site area. The environmental and land use effects from drastic disturbances can be reduced, in some cases nullified, and in certain situations enhanced by timely, well-planned reclamation based on experience and forward-looking research.

Surface mining disturbs about 200,000 acres each year in the USA. Earth construction in building roads, subdivisions, dams, canals, dredging, pipelines, mineral exploration, well sites, and mineral waste disposal disturbs several times more area than does mining. Floods, fires, and other natural events such as the eruption of Mount St. Helens add to the disturbances.

Earth disturbances are often so spectacular that the off-site effects, both positive and negative, are overshadowed. Then disturbance and reclamation become the center of controversy rather than other issues such as need, economics, social and local land use disruptions, and overall environmental effects. Assurance is needed that drastic land disturbances can be reclaimed or at least rehabilitated to avoid long-term pollution or aesthetic problems. We must know how to predict which sites cannot be satisfactorily reclaimed and forecast the impact if they are developed.

High-priority reclamation research needs include: (1) finding ways to reduce the high costs and high energy consumption of reclamation while meeting environmental concerns and future land use requirements, (2) developing methods to use land alternatively and optimize productivity, and (3) validating reclamation performance requirements.

Agriculture and mining supply the basic resources to our economic system. As we as a nation strive toward energy independence, it becomes apparent that we must rely on coal to supply more of our energy in the next several decades. Coal production will be increased through surface mining. In fact, surface mining now accounts for about 60 percent of the coal being mined. More phosphate, iron, sand and gravel, and other minerals are being mined by surface methods. To meet these needs we must strike a three-way balance between protection of the environment, sustained agricultural productivity, and the nation's need for coal and other minerals.

BACKGROUND AND CURRENT SITUATION

The forerunner of reclamation research was work on soil loss and vegetation establishment on severely eroded farm, range, and forest lands in the early 1900's. The focus on drastically disturbed land was sharpened in the 1930's with reforestation research on coal mine spoils. Construction of the interstate highways in the 1950's accelerated research on roadside stabilization and revegetation.

Increased environmental awareness in the 1960's, and more specifically toward surface mining in the Appalachians, led to increased funding for mined land research. Passage of the Surface Mining Control and Reclamation Act of 1977 and the drafting of regulations pointed out gaps in reclamation knowledge that resulted in increased research. The magnitude of this increase in research is reflected in the 3,000 research papers and reclamation management articles listed (Seamalert, 1980) as originating in the late 1970's, as compared with 400 similar articles listed (Frawley, 1971) for the decade ending in 1970.

Reclamation research has often lagged behind reclamation practices used by the better operators. This was because research was geared to then-current reclamation regulations, or because funding was not available for larger scale, forward-looking research. Thus we often settled for working on past disturbances created by environmentally unsound mining or earth construction practices. Reclamation research has tended to be short-term because of demands for fast answers, year-to-year funding, and the pressure on scientists to publish results as quickly as possible.

Four general problem areas are given to provide additional emphasis in this presentation of the background and current situation: Surface Mining, Mineral Wastes, Earth Construction Disturbances, and Natural Disturbances.

Surface Mining

The importance of the domestic mineral industry to our needs, economy, and security cannot be overstated. The mineral industry must continue to explore and expand. Surface mining is an economical, safe method for removing many mineral resources from the earth. As a result, the bulk of domestic ores and coal is surface mined.

The estimated 200,000 acres of land disturbed annually by surface mining in the USA may double by the year 2000. To date an estimated 6 million acres have been surface mined—nearly half of this for coal and about 30 percent for sand, gravel, and stone. One-third of the U.S. coal resources may eventually be surface mined. This represents an area of about 40 million acres. In addition to the actual mining, the associated disturbance by haul roads, handling, and other facilities can double the disturbed area associated with some mines.

Research has been conducted on many site-specific reclamation problems, providing partial bases for laws and regulations relating to mining, construction, and other drastic disturbance activities. Major problems posed by surface mining are disruption of traditional land uses, erosion and sedimentation, subsidence, piping, deterioration in water quality, disruption of aquifers, and exposure of infertile, and sometimes plant-toxic, spoil.

Mineral Wastes

Underground mine refuse and processed mineral wastes such as mill tailings, fly ash, spent oil shale, and coal gasification residues are not as extensive as surface mining disturbances. However, the chemical and/or physical characteristics of these wastes usually pose much more severe disposal, reclamation, and long-term environmental problems. Because of these adverse characteristics, reclamation emphasis is on long-term stability of disposal sites and prevention of water movement through the wastes. In disposing of these wastes, we need to consider that some of them may be reprocessed when more efficient recovery techniques are developed or a market develops for an element not removed in the initial processing.

Huge, black, steep-sided piles of deep coal-mine waste were accepted for years as part of coal mining. Some of these piles supported smoldering fires that produced acrid fumes. Some produced acid drainage. Today, disposal of deep coal mine waste is controlled under the 1977 surface-mining control and reclamation act, and there is a tax on coal currently beng mined to reclaim the old waste piles. But research still is needed to ensure efficient long-term disposal of this waste.

In the past, mine development waste and mill tailings from gold, silver, zinc, copper, and lead ores have been indiscriminantly disposed of, often near or into streams. Some of these wastes can produce acid drainage high in toxic elements. Uranium tailing disposal into unmanaged sites and as fill material has resulted in environmental contamination with low-level radioactivity. Today the disposal of most mill tailings is controlled by state and/or federal regulations. Sometimes tailings are put back into the mine, but they usually are placed into a controlled fill and covered with geological material and/or soil.

Ash (fly ash, bottom ash, boiler slag) from coal-burning power plants poses disposal problems. Depending on the coal source, 5 to 25 percent of the mass of coal when burned remains as ash. About 60 million tons of ash are produced per year and this production will increase with increased coal use. Ash is usually alkaline, and may contain potentially toxic levels of arsenic, boron, cadmium, lead, mercury, molybdenum, and selenium.

About 30 million tons of sludge resulting from lime neutralization of acid from sulfur dioxide scrubber systems is produced annually. This rate may increase to 60 million tons by the year 2000. The sludge is principally gypsum, which might be used commercially, but most of it will probably go into land disposal.

Waste from coal gasification plants will pose disposal problems similar to those of coal ash and spent oil shales. Since this industry is still in the development phase, the characteristics and volume of waste that will be produced are not known. Research is required to characterize the waste, ensure appropriate disposal, and/or explore alternative uses.

Oil shales in the western U.S. contain an estimated 600 billion barrels of recoverable oil—about 100 years' supply at today's use rate. An oil shale industry that supplies 5 percent of the present U.S. oil consumption would produce about 200,000 acre-feet of spent shale per year. Part of the spent shale would be produced in situ and thus, while possibly posing water pollution problems, it should pose fewer environmental problems than the spent shale disposed of in valley fills or built up into mesas. Spent shales are saline and pose plant growth and potential water pollution problems.

Earth Construction Disturbances

The cuts, fills, and borrow areas along the 40,000-mile interstate highway system exemplify many different results of disturbed land reclamation, ranging from successful to unsatisfactory. Progress in erosion control, species selection, vegetation establishment and maintenance, and the blending of disturbed areas with adjacent undisturbed lands has been gradual but consistent through the 25-year construction of the system. Reclamation techniques pioneered by research on the interstate system are now widely used on other disturbed lands.

Today, erosion control and revegetation problems resulting from earth construction disturbances involve primarily dams and water control structures, transportation and energy transmission corridors, subdivisions, industrial sites, timber and mine haul roads, mineral extraction and waste sites, and oil, gas, and other production and exploration drill sites. An estimated 500,000 acres are disturbed annually—twice the area disturbed by surface mining. Improved visual appearance is usually the popular, immediate objective in reclamation of these areas. However, because these disturbances often affect critical sites close to and crossing drainages, this kind of erosion and sediment production must have immediate consideration in reclamation research.

Natural Disturbances

Recent catastrophic volcanic, flood, and fire activity in this country has reminded us of the profound forces involved in natural disturbances. Intensive reclamation measures are often needed to ameliorate the aftermath of these disturbances. This paper does not detail research recommendations for reclaiming "naturally" caused disturbances, although those discussed are generally applicable.

STATE OF THE PROBLEM

Drastically disturbed land situations—surface mining, mineral waste disposal, earth construction—have certain researchable reclamation problems in common and some problems unique to a specific resource or region. Future reclamation, and reclamation research, needs to focus on long-term land use, such as farmlands, forest, pasture and range lands, and/or wildlife habitat. In some situations the use will be more intensive such as recreation or building sites. In disposal of certain mineral wastes, however, long-term surface stability and water and air pollution control must remain the primary long-term as well as short-term objective.

Objectives

In the presentation that follows two prime reclamation research objectives are listed; each objective is followed by suggested research needed to meet that objective.

Objective 1. To devise methods to integrate reclamation practices with earth construction and mining in ways to protect water supplies from adverse effects of drastic land disturbances.

Additional research and technology must be developed to protect water supplies and to minimize damage to aquifers during mining, earth moving and reclamation. Priority research includes studies to:

Reestablish recharge capacity of affected aquifers, improve water infiltration and percolation rates, and characterize spoil and its sequential replacement.

Integrate reclamation practices with earth construction and mining to minimize water pollution while maintaining economically efficient operations.

Define measurements of environmental factors that are critical and standardize monitoring procedures by developing methodology and interpretation techniques for the effects of land disturbances on the chemical and sediment contents of water.

Maximize disposal site stability and minimize water movement through or over the material by developing site selection and final topographic shaping criteria for mineral waste disposal sites. Problem materials are acid-producing materials, uranium tailings, coal ash, and spent oil shales.

Evaluate alternative energy-efficient reclamation techniques for long-term stability and water and air pollution control in mineral waste disposal sites.

Objective 2. To devise practices to optimize stability, productivity, and aesthetics of reclaimed drastically disturbed lands.

Long-term land use and sustained productivity in socially acceptable environments must be achieved in future reclamation. Reclamation can enhance or intensify land use through the development of more-productive soil material, reservoirs, recreation and building sites, biomass farms, wildlife habitat, or landscapes shaped for water harvesting. Research needed to make such reclamation practical follows:

Develop methods to achieve the best use of mined areas considering costs, future use potential, and social and environmental impacts.

Test validity of reclamation regulatory requirements to determine if they enhance or retard long-term site productivity.

Determine the time and inputs required to provide adequate water recharge, nutrient, and microbial levels to achieve self-sustaining ecosystems on reclaimed disturbed materials for areas of permanent vegetation.

Predict site productivity, stability, and visual acceptability based on resource inventory and reclamation planning.

Determine optimum plant succession sequences on disturbed sites keeping in mind the intended land use and plant requirements with a goal of minimizing costs for acceptable reclamation.

Pinpoint the kind and amount of soil, overburden, vegetation, and hydrologie data needed for effective reclamation planning.

Stability problems on disturbed lands such as subsidence, slumping, piping, and wind and water erosion may become apparent during and long after reclamation. Many of these conditions could be prevented or alleviated if appropriate technology were developed and used in planning for the disturbance and for the subsequent reclamation as follows:

Develop improved techniques that minimize erosion and stream sedimentation considering properties of soil and spoil materials, desired compaction, residue management, site preparation, stabilization of materials, and expected storm intensities.

Determine contouring and topsoiling requirements compatible with properties that indicate stability of soil and spoil materials, and that test regulatory requirements.

Determine methods to reduce costs and fuel requirements of final contouring and topsoiling of mined areas and certain other disturbed sites.

Determine if existing milling, burning, or retorting processes can be modified or changed to make mineral waste more suitable for alternative uses or disposal.

Reclamation Resources

Reclamation often emphasizes the negative. For instance, soils or spoils are often too rocky, too infertile, or too acid; slopes are too steep; precipitation is too little or too late. The research challenge is to regard the soils, climate, vegetation, and other factors as resources that will aid in predicting the reclamation potential, and then to use these resources to realize that potential.

Soil. Conservation of the topsoil and replacement after disturbance is a proven reclamation technique on many disturbed sites. Knowledge is needed of the physical, chemical and biological changes in soils with moving, storage, mixing, and their replacement. Research is needed on long-term nutrient requirements and nutrient cycling, particularly nitrogen and phosphorus, and the significance of microorganisms in replaced and mixed soils.

Subsoil, spoil, tailing, waste. Tests are needed that will quickly predict consequences of weathering over long periods to characterize the chemical and physical properties of mineral wastes as plant growth media, and as sources of plant toxins or water pollutants. Of major concern is acid formation and subsequent solubilization of metals in toxic quantities. This has been and is the major problem in revegetation and maintenance of water quality on many mining operations and wastes, as well as on some road cuts and pipelines.

When subsoils or mineral wastes are used as surface or subsurface plant growth media, we need to define the long-term fertility needs, nutrient cycling, micro-organism populations and functions, and weathering. We need to answer the question: "When can this material be treated as a soil?" Part of the answer may be available in research on the genesis, morphology, and classification of reclaimed lands as well as on lands disturbed by floods or sand deposits.

In certain instances, coal mine spoils can be more productive and pose less erosion problems than replaced soils. This is particularly true if clayey acid subsoils are the dominant "topsoil" material. Overburden sampling is costly and time consuming. Research is needed to define required sampling frequency and to find more efficient sampling techniques.

Climate. Climate is a major factor in revegetation of disturbed lands. Limited and unevenly distributed precipitation is often a major problem in the western United States. Reclamation technology must be developed regionally because of the different climate, soils, geology, topography, vegetation, and earthmoving techniques. A research challenge is to consider climate as a resource by defining the more appropriate times for earthmoving, grading, seeding and transplanting. Design criteria are needed for sediment basins and water control and diversion structures in relation to local topography, climate, and soil conditions.

Water. Impounded water can greatly enhance recreational, wildlife, and agricultural uses on reclaimed land as well as on adjacent undisturbed land. However, because of past experiences with incorrectly stored water causing floods and seeping through mine wastes, the practice has been discouraged and even prohibited in some states. Restablishment of aquifers disturbed by coal mining is sometimes required. Research is needed to define if, how, and when aquifers and wet lands disturbed by surface mining can be reestablished.

We need to determine potential chemical pollutants in overburden and mineral wastes and develop methods for preventing pollution of streams and groundwater by these elements. This may involve depth- of-burial studies including burial under non-toxic materials, between layers of them, or by mixing.

Irrigation is becoming a more common practice to establish the desired vegetation on disturbed sites in arid and semi-arid climates. Research is needed on the timing, minimum amounts of water needed, and the quality of water that is acceptable for plant establishment.

Plant materials. Considerable work has been carried out over the past 60 years to select plant species for drastically disturbed sites. This work needs to be continued, with emphasis on selecting species for the more difficult sites such as deserts, and the alpine regions, and for adverse soil conditions such as high acidity or salinity. Selection and breeding within plant species should be emphasized for problem areas. Nitrogen-fixing species deserve continued research emphasis.

A major problem in reclamation is that seed supplies are limited or not commercially available for many species, especially natives, with proven reclamation potential. Research is needed on seed production, harvesting, processing, and storage and preseeding treatment for many native species.

Machinery. Earthmoving may be 70 to 90 percent of modern reclamation costs. A major continuing effort in research and development on more efficient machines to move topsoil and grade spoil is warranted by costs as well as by the need to set up more favorable conditions for water infiltration and plant growth.

Capital investment. Reclamation costs range from several hundred to 10,000 dollars or more per acre. These costs must be justified on the basis of environmental protection, aesthetic and regional land use, because land prices are only a fraction of the reclamation costs. Research is needed to define costs, particularly of reclamation alternatives, that will meet the reclamation objectives. As fuel costs increase, alternatives to the extensive grading now done in mined land reclamation must be developed. The economics of reclamation of mining and mill tailings from minerals such as copper and zinc, which are produced in world-wide competition, need special attention.

Knowledge. Some people regard reclamation as a totally new field with little background to build upon. This is not true. Considerable research in plant and soil sciences, engineering, and watershed management has direct application in reclamation. Much of the earlier reclamation research is in proceedings of symposia or station publications that are now out of print or difficult to find. Even today, much reclamation research information is contained in progress reports that only go as far as the sponsoring agency. Researchers must get this information into forms for reclamation consumers to use.

Scientific journals now accept papers on reclamation research. Even with the new-found acceptability, it is often a question whether a paper on reclamation of a specific waste, produced by a process modified to fit a local ore and under the influence of a specific climate, is of enough interest to publish. Then, publication of a concise note is encouraged. Much know-how gained by field experience is lost when those working in reclamation retire or move to another job. It should be part of organized research to document this knowledge.

Computerized information systems can help with reclamation decisions. The potential is high if adequate adaptive research is conducted. Overall, however, more effective communication and information retrieval are needed.

Planning and management. Reclamation and land use planning should be integrated into the overall earth moving or mining plan. Research is needed on the methods of integration and the necessary detail of resource inventory for successful reclamation planning. Such planning procedures should involve landowners, mining companies, private citizens, and local and state agencies.

Management is "getting the job done." It has often been the weak link in reclamation. One reason is that the mineral resource or the earth structure is considered as the goal and reclamation as an expensive "add-on." Disturbed land revegetation can be regarded as specialized farming or, if you prefer, horticulture or forestry, where timely and appropriate operations such as seedbed preparation and drill-seeding, are of great importance. This is often a difficult concept to integrate with overall earth moving and surface mining operations.

Continuing evaluation of past reclamation research and reclamation practices as a whole is needed. What works? What doesn't work? What are the problems? With this background, some reclamation can be designed and done so it is a walk-away, no-management-needed situation. This approach would be particularly desirable on reclaimed abandoned roads and trails, pipelines, well sites, exploration sites, or other localized disturbances.

Long-term management should be a part of the overall reclamation plan where the disturbed land goes into a well-defined use such as farmland, forestry, pasture, or range. Walk-away reclamation for certain mineral wastes such as radioactive or sulfide tailings, fly ash, or spent oil shales appears unrealistic. Thus, long-term monitoring of disposal sites is needed, particularly of water diversions, groundwater quality, vegetative cover, and erosion. Monitoring results need to be fed back through research so that more efficient disposal can be designed.

RESEARCH PRIORITIES

Drastic land disturbances by mining, mineral-waste disposal, and earth construction greatly alter soil, water, land use, and aesthetics. These alterations, in turn, affect a much larger off-site area. Adverse effects of drastic disturbances on the environment and land use can be reduced, and in some cases nullified, by timely, well-planned reclamation based on experience and research. In certain situations, reclamation can even enhance the environment and land use.

About 200,000 acres in the United States are disturbed each year by surface mining. Waste disposal from mineral processing, along with earth movement involved in roads, subdivisions, dams, canals, dredging, pipelines, mineral exploration, and well sites, disturbs several times as much area as does mining.

Earth disturbances are often so spectacular that the off-site effects, both positive and negative, are overshadowed. Then, disturbance and reclamation become the center of controversy rather than the more basic issues such as product need, economics, and social impacts. Assurance is needed that drastically disturbed land can be reclaimed or at least rehabilitated to avoid long-term pollution or aesthetic problems. We must also be able to predict which sites cannot be satisfactorily reclaimed and forecast the impact if they are disturbed.

Agriculture and mining supply basic resources to our economic system. As we, as a nation, strive for energy independence, it is apparent that we must rely on coal to supply more of our energy. Coal production will be increased through surface mining. In fact, surface mining now accounts for about 60 percent of the coal produced. Also, more phosphate, more iron, and more sand and gravel are being mined by surface methods. To meet these needs we must strike a balance between protection of the environment and the nation's need for coal and other minerals. Disturbed-land research must find ways to reduce reclamation costs, optimize productivity of reclaimed areas, and validate performance requirements. To accomplish these objectives we must:

1. Integrate Reclamation with Mining and Earth Construction

We need to devise methods to integrate reclamation practices with earth construction and mining in ways that minimize aquifer damage during mining and protect water supplies from adverse effects of drastic land disturbances. Technology is needed to achieve the following goals.

a. Restore the recharge capacity of affected aquifers.
b. Integrate reclamation practices with earth construction and mining to minimize water pollution.
c. Identify critical environmental factors and develop standardized monitoring procedures.
d. Evaluate alternative, energy-efficient reclamation techniques for achieving long-term stability and for water and air-pollution control in mineral waste disposal sites. Problem materials are: acid-producing substances, uranium tailings, coal ash, spent oil shales, and others.
e. Evaluate processing techniques to make mineral waste more suitable for alternative uses or disposal

2. Balance Land Disturbances and Environmental Quality

We need to devise reclamation practices that optimize productivity and aesthetics and enhance or intensify land use through development of more productive soils, reservoirs, recreation and building sites, wildlife habitat, or landscapes shaped for water harvesting. To meet these objectives we must:

a. Test validity of reclamation regulatory requirements to determine whether they enhance or retard long-term site productivity.
b. Achieve a self-sustaining ecosystem by providing, for example, adequate water infiltration and recharge, nutrient levels, and microbial populations.
c. Predict and achieve visual acceptability, site productivity and stability based on resource inventory and reclamation planning.
d. Achieve the best use of mined areas by considering costs, productivity, future use potential, and social impacts.

3. Plan Reclamation for Long-Term Site Stability

We need to devise practices for use in planning and reclamation that optimize stability and minimize problems of subsidence, slumping, piping, and wind and water erosion that may become apparent during and long after reclamation, as follows:

a. Develop improved techniques to minimize erosion and stream sedimentation.
b. Evaluate regulations for contouring and topsoiling to determine whether requirements are realistic for achieving stability of soil and spoil material.
c. Reduce costs and fuel requirements of contouring and topsoiling disturbed sites.

REFERENCES

1. Frawley, M.L. 1971. Surface mined areas control and reclamation of environmental damage, a bibliography. Bibliography Series 27, USDI Office of Library Services, Washington, D.C.
2. National Academy of Sciences. 1979. Surface mining of non-coal minerals. A study of mineral mining from the perspective of the Surface Mining Control and Reclamation Act of 1977. National Academy of Sciences, Washington, D.C.
3. Schaller, F.W., and Paul Sutton (eds.) 1978. Reclamation of drastically disturbed lands. American Society of Agronomy, Madison, Wis.
4. Seamalert. 1980. Surface environment and mining alert to current literature. USDI Office of Surface Mining, Div. of Information and Records Managements. Washington, D.C.
5. Soil Conservation Society of America. 1980. Surface Mine Reclamation, a position statement. Supplement to the Journal of Soil and Water Conservation Vol. 35. Available from Soil Conservation Society of America, Ankeny, Iowa.
6. U.S. Department of Agriculture. 1977. The status of land disturbed by surface mining in the United States. USDA SCS-TP-158.
7. U.S. Department of the Interior. 1967. Surface mining and our environment. U.S. Government Printing Office, Washington, D.C.
8. U.S. Congress. 1977. Surface mining control and reclamation act of 1977. Public Law 95-87, 95th Congress.

10

Emery N. Castle, Ralph J. McCracken,
Anson R. Bertrand, R. Neil Sampson,
Dale L. Stansbury

ADDRESSES Plenary Sessions

Land and Water: Natural Resource Use in Agriculture

Emery N. Castle

This workshop is indeed a timely venture. Both agricultural and nonagricultural groups are showing great interest in the natural resources used in agriculture. There is concern about the loss of agricultural land to nonagricultural uses, about erosion, and about water use. Many nonagricultural interests are also aware of such things as declining ground water tables and saline soils. The opportunity exists to channel these concerns toward constructive social action.

An opposing point of view can be advanced. There is great pressure to reduce public spending of all kinds, especially at the federal level. Furthermore, less government, rather than more, is the order of the day and, if social or group action on soil and water conservation means more spending or more government, it may turn out that this is just not possible at this time. Nevertheless, as noted, there is widespread interest and concern about these matters and I sense there is also substantial concern about inadequate public support for agricultural research at the federal level based on a widespread belief that support for such research has not kept pace either with need or with support for research in other areas.

Thus, the challenge to this conference is clear. It is to design a program of research that will be responsive to social concern but which will recognize the budget realities of our time. It is my belief that a document or statement coming from this conference which has the appearance of a laundry list of research projects will not have much influence. But if a statement is forthcoming which reflects research needs for valid areas of concern with careful attention to the relative importance of different kinds of research, it could well prove to be a landmark undertaking. I am not convinced that an obviously self-serving statement will carry great weight in important places.

It is my understanding that the policy panel for this workshop is contrasting the social environment which now exists with that which prevailed during the 1930's. It was during the 1930's that significant social action occurred designed to address the soil and water conservation problem in agriculture. The panel is to be commended for their approach and I look forward to seeing the results. I regret they were not available when these remarks were prepared but, since they were not, it was necessary for me to anticipate their results somewhat. I draw attention to two factors which I believe influence the contemporary setting for land and water management in agriculture in a most significant way.

This country has long followed a policy of abundance with respect to agricultural inputs. We see this reflected in the Homestead Act, the Reclamation Program, credit programs, and in research and education. It was the Jeffersonian ideal that agriculture provided opportunity for entrepreneurship and for the individual proprietor. The abundance of land, water, and other natural resources supported this philosophy nicely. Even when these resources were more fully utilized many social benefits still resulted from these policies. Lower cost food and the release of human resources from agriculture production so that the production of other goods and services were possible, were major benefits. Until recently, even when land and water were fully utilized, their use outside agriculture was such that the opportunity cost was either zero or quite low—less than their use in agriculture.

This situation is different now. Much agricultural land has been worth more in other uses and a great deal has been converted to other uses. The same kind of competition exists for water. We know that agricultural water rights are often desired so that they may be used for municipal or industrial use or for energy development. This is a fundamental change and it influences the kind of information, and consequently research, which is of value.

The other fundamental change is the way agriculture is now viewed as contrasted to an earlier period. When the majority of the population either lived on the farm or had farm ties, agriculture was viewed largely through the eyes of the farmer. Many in the population now have a concern for agriculture only in the context of their food supply. If farmers and agricultural interests wish to obtain the support of these people for research they will need to address problems in the context of this concern. If agricultural exports are going to be encouraged as a matter of national policy I believe sometime during the 1980's the administration in power in this country will have to reassure the consuming public with respect to both the adequacy and cost of their food supply.

Another manifestation of this shift in attitude is the concern expressed by many environmentalists about agriculture. Agriculturalists have a bad habit of lumping all environmentalists together and generalizing about them. This is inaccurate and can lead to much mischief. For example, there is a group which can be identified as those who are concerned about what the economist might call "negative externalities" from agriculture production. This is the damage that occurs off the farm from what occurs on the farm. The costs resulting from siltation, the loss of nitrogen and phosphorus, and the escape of herbicides and insecticides provide examples. There are others who view the problem in an entirely different way and believe that modern agriculture is ecologically unsound and not sustainable in a long-run context. The policy positions and the concerns of the two groups are very different. The list is much longer than just two but two is all that is necessary to make my point that great heterogeneity exists within the environmental movement.

Agricultural interests generally and agricultural research interests in particular need to recognize that they have little political strength acting alone. In order to obtain public support for agricultural research, issues must be addressed that are of interest not only to the farmer but to others as well. There has been, in my judgment, much misguided discussion in the last few years about research and educational support for commercial agriculture. Statements are frequently made that food and nutrition research, or research that affects any nonfarm interest, is always competitive with production research. While this may be true in some instances and in the short run, I am not persuaded that this is true generally or in a longer run setting.

My proposal to you, then, is that this workshop identify research that has the prospect of making a difference in the lives of people. The viewpoint adopted for specifying needed research will need to be broader than that of the individual farmer if the necessary support for the resulting agenda is to develop.

WATER AND AGRICULTURE

One of the major problems facing this workshop will be how to identify major research areas while recognizing that conditions and problems vary greatly from one geographic area to another. We are a large nation and both natural and social conditions change from place to place. In order to make my remarks intelligent it was necessary that I categorize and classify the major problems pertaining to soil and water conservation. I know that several participants have already given a lot of thought to this problem and there has been a great deal of discussion about it. Therefore, the classification that I will use should be viewed as necessary for my purposes but will probably be too crude for the detailed discussion, work, and reporting of this workshop.

I proceed on the assumption that there will be increased demand for both agricultural land and water resources which will come from two sources. One will come from within agriculture itself resulting from the increased demand and higher real prices for agricultural products at least through the remainder of this century. Furthermore, our monetary and fiscal policies have stimulated farmer demand for land and water resources. The other source of increased demand will come from outside agriculture because land and water resources are needed for an expanding economy.

When water and agriculture are discussed there is a tendency to assume that irrigation is either the only or the major water management problem in agriculture. This is far from true. In almost every part of the United States water is either inadequate or overabundant at certain times of the year. The choice of crops, tillage methods employed, and the selection of machinery are all means of adjusting to or managing the moisture supply. Research related to managing the moisture supply is just as much water management as is irrigation.

There are at least three major clusters of irrigation problems in the United States. There are those problems associated with irrigation on established irrigation projects and this set of problems tends to be concentrated in the Western United States which are generally the more arid regions. It is useful to distinguish between surface and ground water because there is a special set of problems and adjustments associated with ground water utilization. The third set of problems are those concurred with increased irrigation. The most rapid growth of irrigation probably will now occur mainly in the humid regions where many of the problems are different from those which exist on established irrigated projects in arid sections of the country. I will discuss each of the three situations establishing the fact that the economic and social questions are different for each.

In 1977, 23 percent of the cropland in the 17 Western states was irrigated with the percentage for the 31 Eastern states being about 4.4 percent. The acreage irrigated in the West amounted to 46.4 million acres while the total acreage irrigated was 9.3 million acres in the East. Thus, most of my remarks will apply to the West but not entirely so. A major problem affecting many of these projects pertains to sustained productivity. Excess salinity is a most significant problem and inadequate or improperly maintained delivery systems is another.

Another major problem for established irrigated areas pertains to nonfarm competition for water. The original water right was usually obtained at no cost and the delivery of water to the farm was often subsidized. The economic value of water is now higher and a host of additional problems are brought to the fore. If water use in agriculture is isolated from the forces of the market, continued inefficient water use in agriculture will be encouraged. There are now many known techniques for improved water use which are unused because it is not in the farmer's interest to use them. If market forces are permitted to come to play in agricultural water allocation, there will be immediate interest in economic means of improving the efficiency of water use. Thus, the research which will be of the greatest value will be influenced by institutional developments.

Declining ground water levels promise to reduce irrigated acreage significantly during the next two decades. No doubt farmer behavior will be much influenced by social policy. In some states permits for ground water usage are not extended if it is believed declining water levels will result. In other states this is not the case and water levels are declining in many areas including the high plains portions of Texas, Oklahoma, Kansas, Colorado, and New Mexico. What are the responsibilities of society if water levels decline? Should water be imported at great cost to preserve the agricultural industry and dependent community services and businesses? Or, should the water level simply be allowed to decline with individual responsibility for adjustment to this state of affairs? It is beyond the scope of this paper to explore all of the ramifications of this problem but some things can be said in passing. The public policy chosen will have a great deal to do with the kind of research that will be of the greatest value. On the policy side, attention should be directed to the inconsistency between resource policies which permit a resource to be exhausted and other policies directed toward sustaining or stabilizing the economy of an area. It may be rational to exploit a natural resource over a particular time period but, if a community bases its existence on the exploitation of a resouce, it may be quite costly to sustain such a community after a resource has been exhausted. If community stability is desired, resource use policy should be shaped accordingly.

The incremental value of a unit of water at the most critical time of a growing season is very high in much of the humid areas of the United States. Although less than five percent of the cropland is irrigated outside of the Western United States, there will continue to be an incentive to supplement natural rainfall in some way. The techniques for accomplishing this as well as the institutions needed for social feasibility, will, in many cases, be quite different than they are for the arid West. In the humid areas irrigation is but one of several ways of increasing output per acre. In much of the arid region crop production is possible only in the presence of irrigation and other farm practices are necessarily associated with irrigation. As this audience knows well, clearing, draining, and land forming activities, as well as irrigation, have been utilized increasingly during the 1970's as a way of expanding and enhancing the cropland base in the humid areas.

Proposed water research should recognize the relation between water institutions and technical agriculture. So long as water use in agriculture is protected from market forces, discoveries of more efficient means of water use may be of little value. But if there were economic incentives for the adoption of such techniques, the value of certain kinds of research would increase substantially.

What happens to the productivity of water in agriculture will have a significant effect on the pressure which is placed on the land base. If irrigated acreage increases on balance, fewer additional acres will need to be brought into production. Thus, the potential of research on water management exists but, a major task facing this workshop is to assess the prospects of success of such research. Potential and actual pay-off are not the same thing.

SOIL CONSERVATION

There may be a greater concern about soil erosion among people informed about agriculture than at any time since the 1930's. There is the belief that many conservation practices have been abandoned, that erosion has increased, and that additional acreage will be brought into cropland which will be subject to erosion. It is most appropriate that this workshop address the problem.

It is known that some practices and some crops are more soil conserving than those that are being applied. Thus, it might be said it is not technical production information that is lacking, but rather the need is to understand farmer behavior and to design measures to change that behavior if it is decided that different behavior is desirable. But how is one to decide that? What is the most socially desirable rate of soil loss? There are two parts to such a question. One part is concerned with the relation of soil loss to productivity which is most certainly a technical question. The other part must address how to weigh future values relative to the present. If one knew the answer to this question we might be able to determine a socially desirable rate of soil loss. But farmer actions would have to be brought into line if such information is to be of practical significance. This requires knowledge of farmer behavior and the role of incentives in influencing that behavior. These questions are not new. They were being asked when I studied crops, soils, and economics as an undergraduate more than 30 years ago. Some progress has been made in obtaining answers but the trouble is that the answers change even though the questions remain the same. That is, the productivity of soils depends on a lot of things, including technology, in addition to the depth of the topsoil and, as time passes, technology changes also. Furthermore, farmer behavior changes in response to the social environment in which farmers must operate and the social environment is very different today than it was 30 years ago. Some of the important differences are: (1) much larger equipment which makes specialization in both crop and livestock production more profitable, (2) an expanding market for farm output in contract to surpluses, and (3) an inflationary economy which places a premium on holding real estate debt.

I doubt seriously that society will wait before taking action until research provides for agreement among research scientists as to a socially desirable rate of soil loss. The "permissible" rate of soil loss that we now have may be regarded as sufficient for policy purposes with concern for ways of bringing farmer behavior into conformance with this rate. The principal arguments will then be about the relative merits of direct regulation versus financial inducements and voluntary versus involuntary compliance. Research can assist in resolving such questions but it is not the type of research most of us are accustomed to conducting. For one thing, it is not in the exclusive domain of either natural or social science but rather calls for a combination of the two which is another challenge this workshop must face. Will natural and social scientists be able to jointly identify important research problems and approaches? Or, are the stereotypes that each has of the other likely to dominate, with the result that problems are identified that fit neatly into disciplinary boxes which will then yield research results that will have to search for a social problem to which they may be applied?

AGRICULTURAL LAND

There undoubtedly will be numerous questions raised at this conference about the adequacy of agricultural land for crop production and whether research has a role to play in this respect. I offer here a few remarks to help place the matter in perspective. In doing so, I draw heavily on the research of Pierre Crosson of Resources for the Future. William Larson has also written authoritatively about the subject.

There are, of course, many factors that influence the acres in crops in the United States at any particular time. The amount of land leaving agriculture and put to non-agricultural uses, the trend in exports, yields per acre (which depend on the kinds and amount of technology), and the amount and kind of land in the less intensive uses such as pasture, range, and forests are all important. It is the opinion of Crosson that additional acreage will be brought into cropland between now and the end of the century. He foresees increased exports and some loss of cropland to nonagricultural uses. He does believe increases in yields per acre are unlikely to produce sufficient agricultural output unless cropland acreage is also increased.

He believes the economic costs of land are likely to increase in the future. If he is correct, then research which would permit other inputs to be substituted for land would be valuable. It is not clear to me that this conference should necessarily emphasize yield increasing research because this area of research has received considerable attention in other meetings, but it is well to keep in mind the relation of such research to the subject under discussion.

Crosson does not believe that the environment cost of bringing the additional land into crop production will necessarily rise, but it may well do so. He bases this conclusion to a considerable extent on the assumption that conservation tillage will increase rapidly in the coming years. He has taken a careful and thorough look at conservation tillage in a manuscript on this subject (Crosson, Pierre. 1981. Conservation and conservation tillage: A comprehensive assessment. Soil Conservation Society of America. Ankeny, Iowa). Crosson has reviewed the literature and has highlighted gaps in our knowledge about the practice.

GRAZING AND PASTURE LAND INCLUDING THE PUBLIC RANGES

At one time in my career I did research on range economics. My recent research has not been on this subject and I must admit to being out-of-date. At one time there was a most substantial gap between the recommunications of range scientists and what ranchers were actually doing. It was never clear to me why such a large gap existed. Since that time other uses of range have come to the fore and there are nonagricultural interests who would influence the combination of uses on the public range lands.

Thus, it is clear there are quite important public policy issues associated with the use of the public range lands. This forms the base of the sagebrush rebellion and other controversial issues associated with these lands. While these are very interesting questions it is not clear to me just how they should be factored into the deliberations of this conference, if at all.

Nevertheless, it is most important that the effect of agricultural research on the productivity of pasture and range land be considered. Only in this way can one decide on research priorities between cropland and pasture land. I encourage this workshop to debate this issue because those who must make decisions about funding research have a right to know what scientists believe about possible pay-off. It is my impression that we know considerably more about past pay-off to investment in agricultural research on cropland than we do on pasture and rangeland. However, this is an impression only and a careful examination of the facts might dispel such an impression.

FOREST LAND

Space and time do not permit me to develop the many problems of forest land use. Nevertheless, it is important that such matters be considered. Again, those who finance research have a right to know about relative returns to agricultural and forestry research. It is my impression that forestry culture is being modified greatly by the application of science to its production. If this is so, it may be the prospects for pay-off are quite good. Again, this is the kind of information that scientists should generate. While it would be expecting too much to ask this workshop to answer such questions, some recommendations could be made as to how such information might be generated.

AGRICULTURE AND THE NATION'S ECONOMY

There is a story about John F. Kennedy, Kenneth Galbraith, and agriculture during the 1960 presidential campaign. JFK is supposed to have said to Galbraith "Ken, you are my agricultural advisor. I want to hear about agriculture only from you. And, come to think of it, I don't want to hear anything about agriculture from you, either." Whether the story is true or not, agricultural matters have been discussed but little in presidential politics since that time. Russian grain sales during the Butz administration in the USDA and the recent grain embargoes are major exceptions.

But this may change despite the declining political power of farmers. If exports increase and retail food costs continue to rise, there will be a great temptation to assert a cause and effect relationship. Since everyone must consume food, consumers, rather than farmers, are likely to be the group to which a political appeal will be made. Also, food exports are going to be viewed as instruments of foreign policy. Thus, agriculture will be important to three issues of central political concern—the cost of living, the balance of payments, and foreign relations.

It would be a mistake to assume that agriculture will necessarily be better understood and that more enlightened agricultural policies will result. The interests which are likely to be dominant will not be concerned primarily with production processes or farmers. This argues strongly for the building of bridges by agriculturalists with the numerous interest groups who will have a concern about agriculture. The inevitable result will be the involvement of many who have little substantive knowledge about what is occurring on the land with food production.

We can also expect that national and international developments will have a great impact on the social environment within which agriculture must operate. The demand for U.S. agricultural products has expanded enormously because of the growth and development around the world and these trends can be expected to continue. But it is also true that the recent performance of the U.S. economy has had a negative impact on agriculture as well. The inflation which has occurred has stimulated greatly the demand for agricultural land. To this time, farmers themselves have been the principal source of this increased demand because land has been such a good investment for them. The result has been a substantial incentive for increased use of land. If inflation is slowed we can expect some economic hardship to be experienced by farmers who have purchased land on the assumption that the inflation rate of the recent past would continue. But if the inflation rate is reduced and if farmers are convinced that the increase will not be resumed, I predict they will favor land-conserving relative to land-using practices in the future much more than they have in the past.

In summary, happenings outside the agricultural industry will have great impact on happenings inside that industry.

Resources Conservation Act and Future Soil and Water Research Priorities

Ralph J. McCracken

I welcome the chance to meet with you today and to discuss some important research needs that were identified through the Department's work with the Soil and Water Resources Conservation Act of 1977, or RCA. But, first, I'd like to spend a few minutes outlining the background and the status of RCA, and what the prospects or directions for natural resource conservation may be in the next few years.

The Act came about because Congress felt it was time to take a hard, new look at the ability of our country's natural resources to meet future food and fiber needs—and to re-evaluate the department's 34 soil and water conservation programs.

For the past 3 years, nine USDA agencies and two White House offices (the President's Council on Environmental Quality and the Office of Management and Budget) have struggled to meet the requirements of this act, which, basically, are to:

- Appraise on a continuing basis the soil, water, and related resources of the Nation's non-Federal land;
- Develop a program for furthering the conservation, protection, and enhancement of these resources;
- Report to the Congress and the Public in 1980;
- Provide Congress with annual evaluation reports; and
- Repeat the process in 1985. (Without further Congressional action, the law terminates in that year.)

Public opinion has played an important role in the whole RCA process. Some 9,000 local and 18 regional public meetings were held, a Lou Harris public opinion survey was conducted, and a 60-day public review period of the four RCA draft documents was carried out during which the Department received more than 65,000 written responses. The draft documents, which were published in January 1980, contain a wealth of resource data and provide the basis for our zeroing in on certain research needs. I will go into this further in a moment.

All of the information gleaned from the public participation effort was worked into the proposed program recommendations, which I was presented to former Secretary of Agriculture Bob Bergland before the inauguration. Secretary John Block now is reviewing them. Once the new Administration decides what form the proposed RCA program should take, it will be prepared and there will be another 60-day public review period.

We believe the overall objective of the RCA program will be to assure the natural resources needed by future generations of Americans for food and fiber, environmental quality, water and energy, fish and wildlife, recreation, and community needs. It will acknowledge that our natural resource base has been degrading and that the impacts of this degradation have been masked by technology development which increased food and fiber production by an average of 1.6 percent annually in the past.

In addition to this overall objective, we expect there will be objectives for each of eight major resource areas. These include:

- Control soil erosion, giving emphasis to economic benefits to landowners and to society. We believe such an objective should ultimately lead to nondegradation of soil resources.
- Reduce flood damages to rural communities and to farmland.
- Help States meet standards in their water quality management plans, priority to areas where agricultural nonpoint source pollutants pose a threat to agricultural production and human health.
- Minimize irrigation water use; reduce ground water withdrawals where water tables are falling; conserve water to protect in-stream uses; and optimize the efficiency of water use from natural sources for agriculture.
- Identify high-quality wetlands and provide technical information about them.
- Enable farm and forest producers to work toward energy self-sufficiency using methods that conserve and protect soil resources.
- Increase the use of organic wastes for soil tilth and fertility.
- Reduce downstream flood damages; prevent unnecessary conversion of important farmlands and wetlands to urban uses; and reduce sediment delivery from urban construction sites.

Each of these objectives will have quantitative targets for the year 2030. They also will be supplemented with specific goals for the first 5 years under RCA. Each 5-year goal is designed to move us part of the distance toward eventual attainment of our objectives.

Now let's turn to the issue of research needs.

Our 3 years of scrutinizing—and analyzing—and evaluating for RCA unearthed an amazing amount of facts and figures about our Nation's resource base. Much of these data, which were published in the Draft 1980

RCA Appraisal Report, were provided by SCS's 1977 Natural Resources Inventory.

Among other things, we identified the major soil, water, and related resource problems facing the nation. These include:

- More than a fourth of our cropland— about 110 million acres–is eroding at rates which lower soil productivity.
- Sediment from erosion will continue to be a major source of water pollution.
- Upstream flood damages will increase.
- The cost of producing food and fiber will escalate due to soil erosion, upstream flood damages, and farmland conversion.
- Our ground water supplies will be depleted in some areas as the total demand for water continues to increase.
- There will be an increasing demand for assistance in solving soil and water problems.
- Anticipated increases in demand for food and fiber at home and abroad will create more and more pressure on the land available for growing crops. We predict that by the year 2030 we will have to produce 84 percent more food and fiber than we did in 1977.
- Our supply of good cropland will be further reduced by suburban expansion, highway and airport construction, and other nonfarm uses of land. We currently are losing 3 million acres of agricultural land to such uses each year, including nearly 1 million prime acres. If this trend continues, 44 million acres of prime farmlands will be converted in the next 50 years, which would have a significant impact on U.S. agricultural productivity.

This concern—that our country might not be able to produce the levels of food, fiber, and other materials needed at home and abroad—is what prompted former President Carter to initiate the National Agricultural Lands Study (NALS). A joint effort of USDA, the Council on Environmental Quality, and 10 other Federal agencies, NALS pointed to a number of factors that apparently dampened productivity growth during the last decade, including:

- The rising costs of fuel, fertilizers, and other energy-intensive inputs;
- Less fertile agricultural land available for cropland uses;
- Lack of reserve supplies of water to sustain past growth rates in irrigated agriculture; and
- The loss of natural soil fertility due to erosion or salinization.

What all this means is that after four decades of agricultural surpluses, we must begin thinking in terms of potential scarcity.

Because of its impact on agricultural productivity—not to mention water quality, flood damage, and fish and wildlife habitat—soil erosion—and the most cost-effective means of treating it—were given top priority in developing the RCA program recommendations.

We were working toward a program that would give major—if not primary—emphasis to such conservation practices as minimum tillage, no till, and similar soil management techniques. Mechanical means for controlling erosion, such as terraces, diversions, and structures, would be given less emphasis in the RCA program, because they're not as cost-effective in reducing erosion at the source.

Such a program would pose several questions for soil and water researchers:

- How effective would the program be in solving persistent erosion problems?
- How would these soil management techniques impact the resource base? Would there be more soil compaction? More reliance on herbicides and pesticides? What would be the affect on water quality? On hydrology?

This is the type of information we need as the redesigned program goes forward. We also need information on the impact of erosion on the productive capacity of the land...the productive potential of the resource base as a whole...and how to respond to catastrophic conditions that put even greater stress on our land and water resources.

The USDA's Science and Education Administration (SEA)...and their cooperating partners in the Land Grant Universities...already have begun working on some of these research needs, and have formulated some long-range plans. This is encouraging. But if we are to meet the increased demands being placed on our resource base, all of these questions will have to be answered in a timely manner.

So far I've concentrated on research needs that deal with soil quantity and quality. There are, however, many other issues that emerge from the RCA process that need your attention, such as:

- How much emphasis should be given to water conservation?
- Should public policies and funds be utilized to shift irrigation water from water-surplus to water-short areas?
- What conservation measures and programs ought to be available in areas irrigated from ground water supplies that are being depleted due to insufficient recharge?
- How far should we go in reducing water pollution from agricultural sources? What are the most effective and enduring approaches for doing this?
- What measures are we willing to take to protect communities from devastating floods?
- Can we make agriculture energy self-sufficient? Can we do it in cost-effective ways?
- To what extent can we use crops, crop residues, and biomass to deal with the energy crisis? What are the limitations on using these options as we move into an era of increasing demand for food and fiber?
- Would further treatment of wetlands increase agricultural productivity? Would it conserve energy? When should this kind of option be given more attention in meeting anticipated demand? What about our need to protect high-quality fish and wildlife habitat?
- How much should we rely on our grasslands to produce red meat? How much more efficient could these lands become if their condition

was improved and they were managed better?

- What will be the impacts of converting forest lands that are in commercial hardwoods into cropland?
- How do we accomplish all the actions needed to meet the increasing demands on our resource base and still maintain a high-quality environment and the amenities people like to enjoy?
- How much should be done by State and local governments? By public universities? By the private sector? What should be their roles?

And, perhaps most importantly—what are we willing to do as private citizens and through public programs to assure that America has options for the future? Do we have the technical information and skills to do it?

We clearly must do more than we have in the past. America's natural resources, although certainly among the best in the world, do have limits in supply and condition—do have limits in the ways we can treat them.

We don't have an infinite amount of time to sit around twiddling our thumbs, mulling over the answers. We must act soon. For example, the National Agricultural Lands Study predicts that by the year 2000, most, if not all, of our country's available cropland will be under cultivation.

When that's gone, where do we turn? Do we convert wetlands, which provide habitat to fish and wildlife? Do we convert commercial hardwood forests, which supply our country with lumber and fiber products? Do we convert lesser-quality lands that are better suited for urban development—lands that would require irrigation, more chemicals, and more erosion-control measures?

As we develop answers to all these questions through RCA, our challenge is to select those program approaches that are effective, that contribute to economic and social benefits, that are environmentally defensible, and that have public acceptance.

Once the RCA program is in place and working, it will meet the intent of the Congress that the "conduct of programs administered by the Secretary of Agriculture for the conservation of (soil, water and related) resources shall be responsive to the long-term needs of the Nation."

The RCA process provides us the best opportunity we have ever had to learn how to balance the need for long-term productive capacity—a permanent agriculture—with the need for economic efficiency, environmental quality, and social values.

But we need your help. For future soil and water conservation programs to be truly effective, we must have a body of research that permits us to describe and quantify the questions and issues that I raised earlier. We must have research that is broader in scope, more sensitive to long-term impacts on the resource base, and targeted to the highest priority problems and opportunities. We must have research that is conservation-oriented, rather than strictly production-oriented.

Whether or not we get the research we need depends partly on how seriously Congress and the Administration take these needs. How much are they willing to spend to protect our resource base? Some studies have indicated that we will need to spend much more on agricultural research and extension work if we are to check the recent declines in the annual growth rate of agricultural productivity.

And even if we get the funds we need, can the research be done in time? This is a frightening question, but it is one that deserves special attention.

We don't have time to procrastinate. We need research that will allow us to make wise conservation decisions—and we need it soon.

I urge each of you to study the RCA findings and to begin working on the answers to these pressing questions. What you are doing at this workshop is certainly a step in the right direction. The Soil Conservation Service will be happy to work closely with you and to help in any way we can.

Soil and Water Resources: Research Priorities for the Future

Anson R. Bertrand

Thank you for inviting me to take part in this National Soil and Water Resources Workshop. I hope—as you do—that we may all come away from this meeting stimulated by some fresh thinking about the wise management of these precious and diminishing resources.

My remarks this morning will concern USDA's soil and water research priorities. To set the stage, I am going to review the background of this conservation research. Then I propose to share some thoughts with you about the soil and water research agenda we have designed to face the challenges of the 80's and beyond.

Fifty years ago, markets for agricultural goods had begun to expand and farmers needed more information in order to make intelligent production plans. The seeds of the revolution in agricultural production were being sown during the 30's by researchers, even as the country suffered through the Great Depression and farmers endured the tragedies caused by the dust bowl. The lightweight gasoline tractor was just coming into its own, and the breakthrough on hybrid corn was news.

Then, in the 40's, America's farms set food production records through the use of antibiotics, new pesticides, improved fertilizers, mechanized farming, and cheap energy. More research and technology developments carried us through the 50's and 60's, until we saw this country producing almost half of all the world's corn—two-thirds of its soybeans—and more than a tenth of its wheat.

In the period of the 50's through the 70's, we witnessed the greatest migration the world had ever seen, right here within our country. Eighteen million Americans—many displaced by modern technology—moved from farms to cities. During the same period, new technology and cheap fertilizer masked the effects of erosion. Concerns for environmental quality diverted significant resources—scientists, engineers, and dollars—from traditional erosion control research to such projects as water quality and pesticide pollution. Agricultural surpluses caused a decline in support for "production" research, including soil and water conservation, and competition increased dramatically for our limited supply of prime agricultural land.

We find that every day in the United States four square miles of farm lands are shifted to uses other than agriculture. Today there are highways and houses, airports and shopping centers on land that once produced an abundance of grain, timber, forage, vegetables, and fruits.

Right here in Wisconsin, 19 thousand acres of prime agricultural land were lost each year to urban build-up or related uses, in the years 1967 to 1977. At the same time, the acreage of irrigated land doubled—a development we see in a great many states.

Yet the Wisconsin loss is low compared with that of Texas, where 116 thousand acres of agricultural land were converted annually—6 times as much as Wisconsin—or compared with Alabama, where 89 thousand acres—over four and one-half times as much as Wisconsin—were converted, with irrigated land increasing from nearly zero to 347 thousand acres—or Florida, where the loss is more than double that of Texas and more than 12 times as much as in Wisconsin.

It doesn't take more than a glance at these figures to conclude that soil and water resources may well surpass energy as crisis issues before the end of the century.

Indeed, the intensifying efforts to develop domestic energy resources are already beginning to have unintended repercussions on the country's agricultural land base. New coal mines, power plants, synthetic fuel facilities, and other producers of energy compete with agriculture for land, and suggest other consequences—such as increased air pollution and intensified competition for water which could affect our production of food and fiber.

Since land-use issues concern us all, I want to touch upon some of the needs that will be created by these pressures—some possible courses of action in meeting the needs—and some of the projects now underway in the Department of Agriculture in working toward long-range solutions.

As for the primary use of our agricultural lands, we have two options in meeting food and fiber demands. The first is to bring more acres of land into production. The price for this path continues to increase as marginal lands brought into production further deplete our dwindling water supplies, increase runoff and erosion, contaminate existing water supplies, and increase the energy required to manage these lands.

The alternative is to develop technology leading to increased productivity on the existing land resource base. The rate of increase of agricultural productivity in this country was about 2.1 percent annually between 1939 and 1965. The rate of growth since that time has declined to about 1.7 percent annually, and indeed for some crops the rates of growth may already have plateaued.

Managing our resources to check severe soil erosion and depletion of water supplies must have high priority if we are to maintain long-term agricultural productivity. And these resources must be managed in ways consistent with the environmental standards we have established. Much of our research in the past has focused on developing ways to maintain yields in spite of erosion. We have largely ignored the long-term effects of soil erosion on crop productivity and the environmental consequences of soil loss. These are things we now need to consider.

On some Corn Belt acres, we lose two bushels of topsoil for every bushel of corn we produce—and the eroding soil particles may be accompanied by plant nutrients and pesticides. Nationwide, we lose an average of more than five tons of topsoil from each of our cropland acres every year from sheet and rill erosion. This must ultimately affect the capacity of the land to grow crops. We must not export our soil with our grain. If our biggest corn or wheat crop also produces the greatest amount of erosion, we suffer a net loss. Failure to control soil erosion on U.S. farms and ranches could double the cost of producing food and fiber over the next 50 years, without regard to inflation or other factors.

We are in a period of significant change for nutrient cycling, although the magnitude and specifics of that change are not yet clear. Throughout early history, man's ability to support himself on this planet was tied to the natural plant nutrient cycle and the consequent availability of food. In more modern times, we have altered that cycle and added elements necessary for increased carbon fixation.

We have supplemented the system with mined resources from earlier eras. We produce nitrogen fertilizer from natural gas—residues from prehistoric photosynthesis. But gas and oil are relatively finite resources and their price tag is rising at a rapid rate. Their future as fertilizer feedstocks is in question. Phosphate rock is also a relatively finite resource, and other forms of phosphate will be much costlier.

All of this means we must reaffirm the importance of recycling the plant nutrients already in the agricultural ecosystems. We must find ways to minimize losses from the soil through immobilization, leaching, and volatilization, and to fix nitrogen biologically in nonleguminous plants.

Other serious side-effects from intensive agriculture include declining water tables, pesticide residues, and water pollution. Solutions to these problems must also be high on our list of research priorities.

SEA is building a National Soil Erosion Laboratory on land provided by Purdue University. Eight scientists will be assigned there to study the erosion process. Studies will concentrate on:

- All possible combinations of tillage and mulching practices;
- Erosion-controlling chemicals;
- Total farming systems directed at erosion control;
- The accuracy of predicting erosion using computer models;
- The basic mechanics of erosion;
- The impact and effect of raindrops on soil particles and chemicals;
- How soil particles are detached and how they move; and
- The characteristics of soils in relation to their erodibility.

All research will be evaluated and assessed for both cost effectiveness and erosion control efficiency.

In our fertilizer-soil-tillage research, SEA is concentrating on making better use of organic materials with emphasis on residue management for medium-to-large scale operations. The work is being done in Minnesota, Nebraska, Kansas, Colorado, Maryland, and New York.

Adding valuable data to our fertilizer-tillage research is work being done by a team in St. Paul under the leadership of Dr. William E. Larson. This team is developing a computer system that will predict how much crop residue may be safely removed from the land under a variety of soil, weather, and crop conditions, and how much must be retained to protect the soil surface.

A revision of Agriculture Handbook 282, "Predicting Rainfall Erosion Losses: A Guide to Conservation Planning," now numbered AH-537, includes an analysis of irregular slopes, management factors for conservation tillage, and extension of the Universal Soil Loss Equation (USLE) to the Western and Northwestern U.S. and Hawaii. We hope this information will give some badly needed aid to the farmer west of the 100th Meridian. We are continuing our research on the accuracy of the USLE for farmers in the Delta States, and we need to do more work to extend the USLE to other areas.

Irrigated acreage in the United States has almost tripled in the last three decades–from about 20 million acres in 1944 to 58 million acres in 1977. Irrigated agriculture now consumes more than 80 percent of the water used in the nation. About 40 percent of our irrigation comes from ground water, which in many areas is being used faster than it is being recharged. Depletion is particularly serious in the Central and Southern Great Plains, where ground water overdrafts exceed 70 percent even during years of average precipitation.

A serious situation is developing in the South Plains. Thousands of farmers in an area the size of California that stretches from the Texas panhandle to South Dakota transformed those acres into bountiful farmland by drawing on a huge underground formation saturated with water. Known as the Ogallala Aquifer, the geological formation is being depleted so fast that some farmers are being forced to cut their production. Experts tell us that Ogallala water is being "mined" as surely as any depletable mineral.

The situation varies from place to place according to the thickness of the aquifer and the local history of water use. But engineering studies estimate that underground water in some areas, such as western Kansas, may be depleted in three to 20 years.

In SEA we are well along on planning for the Moisture Conservation-Plant Stress Laboratory in the southern Great Plains. This central laboratory, supported by two satellite laboratories, will be staffed by 22 scientists representing various disciplines—soil scientists, geneticists, plant breeders, physiologists, biochemists, and agronomists—and a plant antomist, mathematician, engineer, agicultural meteorologist, and a physicist. This will be the most diverse group of scientific disciplines ever brought to bear on a problem area in agriculture.

The planning process has involved not only the American scientific community, but also scientists from Australia, Israel, and Mexico. When this laboratory is completed it will be the only one of its kind in the world.

Another priority in our research program is the work we are doing on non-point pollution. This effort is directed toward meeting the mandates of Section 208 of the Federal Water Pollution Control Act and the Clean Water Act. At this time—and through 1984—we will be involved in approximately 70 research projects on non-point pollution. We are working as hard and as fast as we can to provide, by the 1985 deadline, the most recent and accurate knowledge on "best management practices" in controlling non-point pollution.

There is work to be done concerning water harvesting—the total picture. We are not using runoff to advantage. We have not been using our water very effectively or efficiently. In some areas, particularly in the arid west, some ground water reservoirs are too saline to grow most food crops, so they are not now utilized. This water could be used to produce salt-tolerant plants—plants that could be used as biomass to produce energy. An area of great potential exists in improving the inventory, analysis, and effective use of our water resources. We must use models and modern data processing techniques more effectively to improve the efficiency of water use.

Another research opportunity is to find a workable system for modified dryland agriculture to make the best use of available moisture, using only limited irrigation. There is some good work being done in this regard at Bushland and at Lubbock, Texas, where scientists have developed integrated agri-land-water management systems that prevent erosion and conserve water.

Conservation tillage is the fastest growing agricultural practice in history. U.S. farmers employed conservation tillage practices last year on some 60 million acres. These practices, including no-till, are proving to be a primary soil conservation measure. But research is needed on weed and insect control techniques for minimum northern Corn Belt.

Our erosion and water shortage problems come at a time when agriculture faces major productivity demands to help meet the nation's fuel needs. In the future, a portion of our energy needs may be met directly by growing "energy crops" on agricultural land.

A great many uncertainties exist about the economic feasibility of biomass energy approaches. Literally tens of millions of acres would have to be dedicated to this form of energy production if biomass energy were to make a major contribution to the nation's total energy effort. And in considering biomass energy production, we must look at the potentially negative effects, including greater soil degradation. This is a problem that could become acute if poor quality land is brought into production to meet the increased demand for feedstocks, or if unacceptably high levels of crop residue removal occur. High use of biomass for fermentation into fuel alcohol would also mean a relative shortage of the organic material vital to the continuation of the nutrient cycle and the maintenance of soil structure.

In dealing with the primary goal of maintaining and increasing food and fiber production on a limited land base, are there perhaps organic materials going unused? A 1978 USDA study reported bluntly "No". Crop residues and animal wastes make up three-fourths of the available organic wastes. We found that 90 percent of the animal wastes were put back onto our fields, along with the 68 percent of the crop residues. We may be able to increase the percentages returned to some degree, and to improve storage and handling for maximum nutrient retention, but there are no major organic wastes from agriculture going unused.

Logging and wood manufacture produce significant wastes, but these are being increasingly recycled back into the industry they came from.

The only major underutilized resource USDA found was sewage sludge. With about one-fourth of this sludge used effectively, its farmland use can be expanded. However, only about 4.4 million tons are produced annually—compared with the 600 million tons of animal wastes and crop residues.

And there are major problems in putting industrial wastes and municipal refuse on our fields in any substantial quantities. Industrial wastes often contain acids, heavy metals or other contaminants posing a danger to soils or crops. Municipal refuse has some organic content, but it also contains a great deal of inorganic material—such as glass, plastic and metals-and is unfit for use— unless the inorganic material is separated out in the waste-handling process.

In other words, there is a growing shortage of good quality organic matter for use in maintaining and improving the productivity of our agricultural soils.

We know that in the long run, biological productivity is governed largely by photosynthetic processes, and the limitations they impose. There is a great deal we do not know, however. Take nitrogen and phosphorus, for example. We do not know how to provide organic nitrogen for crops without keeping a lot of relatively expensive acres growing legumes in rotation. We do not know how to apply inorganic nitrogen efficiently. Studies show that only a quarter to half of the nitrogen applied to our fields is recovered in the harvested crops—the remainder stays in the field in crop residues, or is carried away in run-off water, becomes gaseous and is lost, or moves down through the soil profile where it poses a potential problem to groundwater quality. With the shortage of natural gas and the price of nitrogen fertilizers rising rapidly, and with serious ecological questions yet unanswered, we cannot count on feeding the world in the years ahead simply by dumping huge amounts of inorganic nitrogen on our fields to boost yields.

A better understanding of plant nutrients and their basic relationships in soils and hydrology must be an important part of our immediate and long-term research agenda.

An increasingly important tool in achieving this understanding will be the multidisciplinary scientific team. We are finding that we cannot afford to stay compartmentalized. None of the natural phenomena we study exists in a vacuum. There is constant interaction, and something important is lost by over-specialization.

In the years ahead, this interaction will become more significant. Equally important is the ability to communicate, not only with other scientists, but with extension specialists, and through them with users of research findings.

Crucial in the era we are entering will be the alliance between farmers, the State Agricultural Experiment Station, and the Extension Services County Agents. Also essential will be a degree of planning and cooperation among farmers that may run counter to traditions of individualism and self-reliance on the old frontier. Research has an important role to play in providing data and other information to help farmers sustain good yields under conditions that conserve soil and water and require minimum amounts of external inputs.

Although we have many successes to report and have done some excellent work in the past, we are now at what might be described as a "watershed"—where either there will be increased awareness and concern for land and water, and accelerated conservation of these resources, or we must give up some of our agricultural export programs and be faced with land shortages, higher food prices, and other far-reaching costs affecting not only agricultural productivity, but our way of life as well.

This is not a prediction of the apocalypse, but rather a frank, scientific assessment of the need for more knowledge in the years ahead.

I am optimistic. There is every indication that we have turned a corner in renewing our commitment to basic research. We are on an upswing. This upswing is evident in a number of ways. For one thing, it is heartening to note that in all of the talk about budget reductions in government programs, there has been no talk of lessening allocations for research.

There is little doubt that efforts of many of your organizations have contributed enormously to the upswing. You have all worked hard to raise the consciousness of the people in Washington to the urgency of basic research in natural resource management. Secretary Block realizes that, given the technology we have today, yields are plateauing, costs are increasing, and environmental pollution is continuing. He also realizes that basic research on the chemical and physical properties of soils will provide opportunities to develop farming practices that will maintain our most productive acres and permit more efficient use of less-than-prime lands in the production of food and fiber.

Speaking at the Annual Farm Bureau Convention in New Orleans last month, Secretary Block said that the United States needs expanded agricultural research, and he predicted that the nation would not be able to meet the challenges of the 80's without such research.

We know that we will have to do some careful planning in selecting our priorities and committing our funds and efforts where they will do the most good in meeting the very real needs that lie ahead.

There are no quick and easy answers. Long-term commitments of resources are needed for conservation research. Data that are difficult and tedious to acquire must be collected over a period of years.

We must take into account the need for human resources. Currently there is a severe shortage of agricultural engineers, a shortage which will worsen in the future. And there is a future need for agronomists and soil scientists to meet a long-term commitment to agricultural research.

The needs that we are called upon to meet in the years ahead, beginning right now, call for a mix of basic research with applied and developmental research. We must rely heavily on intensive extension and SCS technology transfer, education and technical assistance programs—all closely integrated—to reach users.

We have a big gap in public perception of agricultural research, extension, and teaching and what they mean to this country. We must all take some responsibility for closing that gap. Somehow we must get the public to perceive a societal need as obvious, serious and urgent. Then we must clearly state how our work contributes to meeting that need. And finally, we must effectively communicate both the need and the required research, extension, and teaching activities to the public, legislative bodies, and State and Federal executives.

We must be prepared to describe research needs and goals in ways that speak to public concerns. Opinionmakers and policy shapers need to know that we can make these kinds of contributions.

Does the public know that the rate of growth of food crops is not keeping pace with needs? Do they know that the rate of increase in production per unit of land on most crops is slowing down and tending to level off? Do they know there cannot be the increase in crop yields they have seen in the past, and other accomplishments, without a greater, more sustained research effort? Do they know that we are losing our prime farm land at an alarming rate?

The future of our food supply depends on public opinion that sees the wisdom of an investment in basic research. We have a powerhouse in our science enterprise, and we have a responsibility to carry the message of our technological strength to help our nation face the important reality that science cannot be left out of coming social changes. And we have an equal responsibility to see that societal needs are not left out of scientific planning.

Today, as we face hefty challenges, with more tight budgets and increasing competition for funds, let us be honest with ourselves in setting goals and objectives that speak to the good of all the people—that meet real and specific needs. We must ask ourselves, "What concerns should we address? What needs can we meet? Are we talking in language that can be understood? Are we believable? Are we talking to the proper audiences with appropriate messages? Are we making the extra efforts that may be required of us to insure rational decisionmaking?"

I intend to use these questions as guides in making the strongest possible case for food and agricultural research, extension, and teaching on every occasion I can and at the highest appropriate levels. I invite you to join me in this effort.

Public Policy Directions for Soil and Water Research

R. Neil Sampson

Agricultural management is really the management of a man-made ecological system. Viewed through the eyes of an ecologist, agriculture can no more escape certain ecological laws than Newton's apple could ignore the law of gravity. Any ecosystem, whether man-made or natural, must in the long run achieve a steady state in regard to both energy and materials. Fertility, organic matter, energy and water must flow into the system in relationship to outflow, or the potential productivity of the system will change. The soil is a living resource reservoir of constantly shifting matter, energy, and productivity capability. If the size or capability of that soil reservoir is steadily shrinking, it is only a matter of time until that becomes a problem to the animals or humans that depend on the production of this soil for their sustenance.

Agriculture and agricultural technology are, to that soil reservoir, like a pump on a well. The amount of food produced can be increased, as can the water from the well, by speeding up or improving the pump. But those "speed-up" efforts will only render an easy profit so long as the level in the reservoir is maintained. If we begin depleting the reservoir, improvements in the pump and ever-increasing energy inputs to run it faster become, instead of net gains, essential responses needed to maintain current levels of production.

American agriculture is in precisely that position today. Past productivity increases have come through the adoption of new technology and an increase in the amount of capital used, during a period when climatic conditions may have been unusually favorable. This has enabled a steadily dwindling number of farmers in America to stay in business by growing larger in size, adopting the latest techniques, and absorbing narrow (or non-existent) profit margins. But how much more of these trends are we able to absorb? Is the agricultural "pump" speeded up about as fast as it will go? All the evidence indicates that the "water level" in the soil reservoir is continuing to drop.

Unless we turn to the land that makes up the reservoir from which agriculture pumps food, and find ways to reverse the damage and loss that is steadily depleting it, serious agricultural problems seem certain, and in the not-too-distant future. And those problems have important social consequences, for they have the effect of passing the costs of today's production on to future generations. These consequences have received little attention in the agricultural research efforts of the past, and whether or not they do in the future is an important policy question that must be addressed today.

We have essentially two ways to focus research. We can concentrate on the plants and animals, breeding varieties that are more and more efficient at extracting minerals and water from the soil and converting them into useful products. That kind of research can make soil virtually unnecessary, since it is completely possible to grow plants in a medium of sterile sand, providing all the water and plant nutrients from artificial sources. Take the technology further, provide protection from adverse climatic conditions through constructon and operation of a greenhouse, and crops can be grown in a completely artificial, subsidized environment. For some high value food crops, that is a possibility. But for all the acres of corn, wheat and soybeans in America, it isn't too rational.

But what of the "mid-range" of that technological approach? What is the effect of breeding varieties of wheat, corn and soybeans that are more efficient at exploiting the soil? This is a technology which may raise agricultural production in the short run, but may actually lower productivity over the longer term. When we look at productivity as the long-term sustainability of productive resources, including soil, water and viable human communities, it is clear that the development of more and more exploitive agricultural technologies is harmful, not helpful.

New agricultural research, if it is to contribute to the long-term strength of the U.S. agricultural plant, must focus on the types of technologies that allow a sustainable agriculture without soil or water depletion. In our analogy of the pump, we must work at ways of keeping the well healthy and productive, as well as find all the ways we can to efficiently use the water we can extract at non-exploitive rates, not just concentrate on methods of turning the pump faster. There is no such thing as strength through resource exhaustion, regardless of the compelling arguments that are sometimes made for such a strategy as a short-term expedient.

The current trends of soil depletion and waste must be turned around. That won't happen by some nostalgic return to some kind of mythical "good old days". It will only happen by an aggressive, innovative search for agricultural technologies that give the production needed while building and maintaining productivity.

That search must be led by the agricultural scientists in USDA, in State Agricultural Experiment Stations, and in other public sector agricultural research facilities. It won't happen in private industry research, because it is not the type of technology that will provide the payoffs needed to fuel private research. And it may not happen in the public research facilities unless they get the kind of public guidance, support, and funding that will be needed to push it in that direction.

Some of the issues that will need to be faced are going to be difficult, there is little doubt. Let me suggest a few for your consideration.

Can we design research approaches that will study the entire ecosystem that is affected? Can agronomic research on new crop varieties, management methods, and tools give answers as to the long-term effects on the soil? Can we identify more of the cause and effect relationships involved in improving or depleting soils? Let me ask a more basic question. Can we identify a useful measurement for soil quality, one simple enough to be used in the field in order to monitor the effect of management changes?

Can we design more flexibility into our cropping systems? What if the climatic conditions in the next decade are more variable than in the past? Have we fully considered the challenge to research in helping farmers be more able to respond to changing — or adverse — conditions?

Can we continue to expand our knowledge of the off-site effects such as water pollution that result from different management methods, and search for feasible ways to minimize them?

There are dozens of questions like that, and I hope this workshop is successful at identifying and prioritizing them. The latter may be by far the most difficult, and important, part of the exercise. I think the near-term future for agricultural research in general, and soil and water conservation research in particular, looks fairly bright. But that does not mean you will get to do every little thing that pops into somebody's head. It will be as important, or more important, right now to have a hard-hitting list of high priority research needs that can be aggressively supported for funding at both the federal and state levels.

Let me switch for a moment to the more political aspects of where we might go from here, assuming that you will be successful in getting a solid list of research needs and documenting the reasons why they must be addressed soon.

In a recent book, Don Paarlberg, former chief USDA economist, does a marvelous job of reviewing the agricultural policy scene, using the metaphor of football. He claims the game of public policy formulation is largely affected by who has the initiative — who is carrying the ball.

Sometime in the 1960's or 70's, Paarlberg argues, agriculture and its supporters lost the ball. The initiative on many issues switched over to other people — consumers, environmentalists, and others. Instead of being able to decide which issues would be on the policy agenda, agriculture was forced into the defensive posture of dealing with the issues of others. The analogy to football is apt. It makes a big difference whether or not you have the ball.

As Paarlberg points out, it is bad to lose the ball. What is worse, however, is to lose it and not know you have lost it! I think agricultural leaders, although they took a long time to realize that the initiative had truly changed, have thoroughly absorbed that lesson. Now we face the same danger in the opposite direction. We may be getting the ball back, and it is important that we recognize the fact and switch our tactics once again to offense.

My reading of the policy scene is that the outlook for agricultural research proposals is brighter than it has been for years, if we take the initiative properly. My discussions with Secretary of Agriculture John Block and his staff indicate that they have three top priorities: (1) agricultural research; (2) soil and water conservation, including agricultural land preservation; and (3) marketing farm products. This meeting is discussing two-thirds of that policy agenda. Not bad. On Capital Hill, I am running into much the same story. Senator Roger Jepsen, chairman of the Senate Agriculture Committee's subcommittee on soil conservation, recently told an Iowa audience that he strongly supports added efforts in soil and water research. In a poll he took of his constituents, 95 percent listed soil erosion as a significant problem and 71 percent favored an expanded conservation research program. Other Members of Congress, on both sides of the political aisle and in both Houses of Congress, echo that sentiment.

The fact that those good feelings toward soil and water conservation exist does not mean that funds will flow easily. As you know, lip service support for conservation is never in short supply in Washington, but money for programs is always limited. What it means to me, however, is that the climate for telling our story is going to be different, and that means we should change the way in which we approach the task. We have the ball again on this topic, and we ought to take it and run as hard as we can.

I think the research community needs to concentrate on several strategic questions first, however. First, can you make a case for the absolute need to be financially independent with your soil and water research programs? In the past, it seems to me, too much of our agricultural research capability has been purchased at bargain prices by commercial and commodity interests. For a small share of the total research effort, those interests were able to establish the priorities on many of the research stations. And those priorities were nearly always aimed at short-term production goals, often at the expense of long-term soil productivity. I don't blame the companies and commodity groups for wanting this type of research — their customers and members demand it. And I don't blame the research stations for cooperating with them — their constituents, the farmers, stood to benefit, and often budgets were so tight that outside contributions were sorely needed.

But soil and water conservation research priorities are only going to get attention when the research agenda is controlled by the public, and that means when the budgets are solid enough to have that portion of the research program funded by adequate public appropriations. That must be our goal.

How can you shorten the time between the laboratory and the field? What kinds of approaches are needed to get farmers to field test new systems? How can you harness the tremendous innovative capability of farmers themselves, in adapting systems to the soil and modifying them to make them more practical, or better adapted? How much of what we are needing is not so much research as it is technology transfer, and how can that be accelerated?

How do you intend to deal with the institutional and social questions that are so important in soil and water conservation? Often it is not that we don't know how to manage soil properly, but that we lack ways to provide the right kinds of incentives or that farmers will apply what is already common knowledge. That might be called "tactical research". Nowhere near enough of it is getting done today. How do you propose to remedy that?

How do you propose to use cooperating agencies and organizations as you move forward? You need to design joint ventures at the field level that can involve conservation districts more closely, for example. When they are part of what you are doing, and feel you are doing it to help make their local programs more effective, they will help support your programs with far more vigor than when they are simply spectators.

Finally, we all need to work to tell the story to the public. In case you haven't noticed, that is happening today at a pace unparalleled in my experience. There have been many good newspaper stories and magazine articles in recent months and there are several more in preparation at this time, along with at least three books of which I am aware. With the Resources Planning Act, the Soil and Water Resources Conservation Act, and the National Agricultural Lands Study results now available, there is a much better basis of facts to support the conservation story today than there has been in the past.

What this will mean in terms of added public understanding and support is hard to say, but it is difficult to see how it can hurt. The best advice right now, I would say, is that each of you individually, and each of your organizations, should be cognizant of the fact that our story is "selling well" right at the moment, and we should be aggressive in taking advantage of that.

Finally, I think we need to talk about how we can get someone working full time in Washington to tell Congress and the Administration about the research needs for soil and water conservation. We in NACD try to include research in the work that we do, but we have far too many issues to cover to give it the attention I feel it needs. I don't know what the answer might be, but we would be happy to explore that question further with any of the organizations here that might be interested.

Politics and Pressure Groups at the Federal Level: Turning Words into Action

Dale L. Stansbury

Henry Ahlgren, an Agronomist, wrote in the 1948 Yearbook of Agriculture, "The soil comes first, it is the basis, the foundation of farming. Without it, nothing; with good soil, good farming and living."

Not only do soil and water resources undergird our remarkable agricultural industry, agriculture is the foundation of our total economy. Unfortunately, through much of our early history, little thought was given to the preservation of these fundamental resources.

The seemingly limitless frontier, and a lingering hatred of European tenancy restraints contributed to a gullied and wasted landscape and the infamy of the Dust Bowl.

This is not to say that we haven't had individuals who were concerned or that there were no national policies to redress the mindless exploitation of soils and water. For the past century, there has been an increasing consciousness of resource conservation. Teddy Roosevelt was a champion of conservation and raised it to a national issue. Subsequently, soil reclamation and reforestation efforts began to have effects. In the 1940 Yearbook of Agriculture, H. H. Bennett, the Chief of Soil Conservation Service, declared, "Our soil can be saved and soil conservation is practical for the United States."

In spite of increasing awareness and a national policy for soil and water conservation going back at least four decades, we are confronted with extensive documentation that our soil and water resources are deteriorating at intolerable rates.

The U.S. Department of Agriculture, responding to requirements of the Soil and Water Resources Conservation Act of 1977, found the following:

• Anticipated increases in demand for food and fiber at home and abroad will create more and more pressure on the land available for growing crops.

• Our supply of good cropland will be reduced by suburban expansion, highway and airport construction, and other nonfarm uses of land.

• More than a fourth of our cropland—about 110 million acres—is eroding at a rate which lowers soil productivity.

• Accelerated erosion is taking place on millions of acres of forest land and on pasture and range.

• In some parts of the country, water withdrawals are exceeding water recharge year after year, and a great deal of irrigation water is lost before it ever reaches the crops. In addition, streams and lakes are being polluted by runoff, which often includes pesticides, fertilizers, and organic wastes as well as sediment.

• Flood damage, partly as a result of increased storm runoff from urban expansion and construction, will increase during the decades ahead unless something is done.

• Further, the nation will continue to lose wetlands to farming and urbanization at the rate of 300,000 acres a year.

The Global 2000 report reiterates these findings and raises questions about our ability to meet domestic and international food, fiber, and energy demands over the next few decades.

The essence of every soil and water resource assessment is that we face major problems. One of the most serious of the problems is we lack a reliable data base and analytical capability to deal with the problems and decisions that must be faced.

This lack of research base is not just limited to the physical questions of soils and water but is especially weak in social and political areas. Very little specific information is available about the factors that influence farmer decisions in resource management. We have not yet integrated the effects of technology change, population growth, export expansion, biomass demand, or farmland conversion into our soil and water conservation efforts.

Our beginning point is that our soil and water resources are endangered by current demand and practice pressures. This in turn endangers our agricultural industry and the national economy. We have determined that new policies are needed and that extensive research and extension efforts must be undertaken to forestall the demise of our resources. All of these conclusions are supported by empirical evidence and scientific logic.

Clearly, we need an improved set of soil and water conservation policies and programs. It is also apparent that we need research to support these efforts. The problem is how can ideas and words be changed into funding, programs and action. Or in other words, how can we turn a scientific imperative into political reality.

First, let's look at the word politic. It comes from the Greek and means, "relating to the citizen." My dictionary includes four definitions that I'll somewhat abbreviate:

(a) Wise or prudent;
(b) Well devised or expedient;
(c) Crafty or unscrupulous; and
(d) Constituting the citizens.

For many people politics begin and end with the idea of unscrupulous, however, I would argue the body politic includes all of our social goods as well. In order to establish a national soil and water research program that meets national needs and that addresses national priorities, you must work within the political system.

To achieve results within the system, you must understand its parts and how they work and relate. Further, you must devise a strategy that clearly delineates your objectives, identifies focus, and is compatible with the system you are trying to influence. In other words, you must be prudently crafty in advancing a well devised plan and make it meaningful to the diverse citizenry who are interested in various aspects of soil and water resources. You must also convince some disinterested people that they have a stake in soil and water conservation.

Your principal objective is to achieve expanded support of important soil and water resources research. There are several cross currents in this workshop but money is the main stream issue. This is based on the premise that better data and technology can support improved resource conservation.

Fine, but how do we do it?

What I am going to do is suggest three focus groups and discuss some approaches that should be considered. The three groups are the general public, interest groups and the Federal establishment. These are not neatly devisable but for purposes of discussion let's separate them.

With regard to the general public, we are starting with the advantage that nearly everyone agrees that we should protect our natural resources. We have had something of a soil and water conservation policy for decades. The dilemma is implementation, support, and interests are highly fragmented. To resolve this, I would suggest the issue of soil and water resources must be changed from a motherhood issue to a bread and butter issue.

Let me try to illustrate this perception problem. Everyone agrees with the policy statement that soil and water resources should be conserved. Unfortunately, the same people are seldom in agreement when actual programs are implemented.

Policy statements tend to deal in generalities that permit each advocate to conjure up their own idea of "good policy and programs". For instance when someone says wetland conservation practices, a farmer may think about channelization of wetlands and turning them into an alfalfa field, whereas an environmentalist is thinking of it as a reserve for cat-o-nine-tails and wood ducks, but no farmers. Still, they are happy with their ideas. However, when that policy statement is turned into the program, be it the channelization or reservation for ducks, we have conflict.

As long as the conflict is over a narrow issue, be it a haven for wood ducks or a farmer's right to make money, it is not a national issue. Only the opposite vested interests are involved. I believe that the importance of soil and water resources to our national economy must be established. The citizens must understand that they have a stake. Thus, I would suggest that an educational effort is needed. Stop talking just to scientists and start talking to real people. Further, you must speak in terms and in a context that is understandable to people.

This is a public awareness effort. The value will be a general improvement in the political environment for your effort. I don't believe that public support will ever be great enough to ensure funding but it can be cultivated to prevent an outcry against resource research.

There are a large number of interest groups in the soil and water research area. I believe that a major effort must be mounted to establish supportive working relations with such groups. This is important to achieve greater support but also to avoid unnecessary conflict.

The first order of business should be to form allegiances with other agriculture research, extension, and teaching groups. Because of the discipline and departmental organization of universities there is often fragmentation of effort. For example, forestry scientists don't always work as closely with range scientists as they might. Neither works as closely with the general soil sciences groups or plant sciences as they should. This is especially true when it comes to soliciting funds.

An axiom that I used when I worked in the Senate was, "Scientific objectivity disappears in the face of a dollar sign".

Too frequently, one group of Agricultural reseachers argues that another set of agricultural scientists is getting an unfair share. A few years ago an agronomist testified before the Senate Agriculture Committee that the World food crisis was because too much research money went to animal research. You've all seen the rebuttal in the Animal Agriculture report.

This national workshop was convened partially to address this problem. It is pulling all the resource people together. However, you must be careful that this effort doesn't result in new divisiveness at a higher level. At least part of the stimulus for a national workshop for soil and water resources research priorities is the success of a similar effort by the plant sciences some years ago and the animal agriculture effort last year. I hope that this isn't just an effort to achieve an equal voice with plant sciences and animal agriculture.

You must strike a careful balance of telling the importance of soil and water resource research and extension without threatening other agricultural research and extension efforts. Concentrate on the complementarities and I believe that moderation and cooperation can be fruitful through expanded allegiances. Agriculture has too small a political base to tolerate division and competition among its various sectors, especially, when such division is not warranted or needed.

I also want to caution against divisions among research, extension, and teaching. We have a marvelous system but like any human institution there are some weaknesses. Any strategy that you advance must reduce such weaknesses, not increase them. This effort can have dual payoffs. First, it will provide broader support in the Federal arena but equally important it could give you better understanding on campus. Hatch funds are discretionary and are allocated on the basis of the "political" pressures of each campus. If you are successful in showing that your work is complementary to agronomy, animal science and rural sociology you could fare a bit better. Don't go to the bank yet but give it a try or two.

Beyond your scientific counterparts there is a host of soil and water resource interest groups. The interest groups tend to be narrow and clear in objective. Under the umbrella of soil and water conservation you can find the Limestone Institute and Land Improvement Contractors whose principal interest is selling products or services—good old fashioned greed. The Soil and Water Conservation Districts and farm organizations have real concerns about resources but it is couched in a production context of costs and benefits. There are also the environmental groups as well as wildlife management groups whose hearts are great but sometimes questions are raised about their heads.

In many cases, these groups are more notable for differences than similarities. For example, the environmentalists and the Land Improvement Contractors are usually diametrically opposed. Some are very narrow such as the Limestone Institute which focuses almost exclusively on the ACP liming program. However, each is working to influence soil and water policies and programs.

I believe the universities have a unique opportunity within this system to not only tell the story of research and extension, but to serve all of the actors. Because of their involvement in the national arena they can be very supportive of your effort—if they know your story and you have credibility.

One last comment on building allegiances. There are active interest groups that don't look to be logical allies but, if you tell the story in the right context they could be. For instance, Bread for the World is an active world hunger effort whose principal concern is the expanded flow of food to less developed nations. However, if they understand that the loss of soil and water resources will endanger this flow I believe you can count on them to support soil and water conservation.

Another group that you should form a working relationship with is the international development group. If we seriously want to manage the pressure on our soil and water we should work to develop food production capacity in the third world. At this time, we have a total dedication to export expansion. Most commodity groups oppose development of competing supplies. However, the long-term interest of our resource base as well as world peace could be served through agricultural development in the LDC's.

There are two reasons for working with other groups. The first is to avoid infighting and the second is to achieve broader support. Many of these interest groups have contacts, credibility, and support with constituents and government agencies that you can't reach. If you can convince them that your goals are good they can give you significant aid.

I now want to look at the labyrinth called Washington, D.C.

There is no simple way to explain the maze of offices, agencies, interest groups and people that make up the Federal establishment. Frequently, the total is far less than the sum of the parts but on occasion the total is greater than the sum of the parts.

There are three principal groupings—the Congress, the Executive, and the interest groups. Each of these is tied to the others in formal and nonformal ways. Each initiates as well as reacts to initiatives of others. There are constantly changing lineups but there are also some unique traits in each. I've already discussed the interest groups so I'll focus on the Congress and the Executive.

The Congress is 535 individual stories. As a group they have limited expertise but have the greatest flexibility and are dedicated to good. Senators usually exhibit broader interests than Representatives because of the broader constituencies, but both houses must be convinced to achieve support.

To win support, it is helpful, to find at least a few members in each house to be spear carriers. To be most effective, it is desirable to find supporters who serve on the critical committees. The committees of greatest concern are the Appropriations Committees, the two Agriculture Committees and the Environmental or Resources Committees of the respective houses.

In working with the Congress it is also necessary to work with Congressional staff. The committee staffers are of greatest consequence, but individual office staff should not be overlooked. In fact, the various universities should make efforts to establish a good relationship with the appropriate staff people for their State delegations. Whereas, the broader workshop effort could focus on the spear carriers and committee staffers. The Congress has several support agencies; GAO, Congressional Research Service, Office of Technology Assessment and the Congressional Budget Office. Each of these organizations has people in the soil and water resources area and Congress uses them extensively. It is therefore a good idea to identify the soil and water resource people in the support agencies and to work closely with them.

The Executive branch is more formally organized and has greater internal expertise than does the Congress. However, the formality of organization sometimes prevents cooperation among the various agencies and departments. Therefore, it is not adequate to concentrate on only your counterpart agency in the Executive branch. At a minimum you must work with SEA, SCS, Forest Service, and ASCS in the Department of Agriculture, Bureau of Land Management, Geological Survey and Fish and Wildlife Service in the Department of Interior, the Environmental Protection Agency, and the Army Corps of Engineers.

Each of these have unique mandates and agendas, but all affect soil and water resources. There is a need for a bridge to connect these diverse efforts and the university community could provide it.

The land grant universities, experiment stations, and extension services were chartered by the Federal government and have a continuing relationship through formula funding. This provides access to the agencies and the Congress. Since the responsibility of the universities is for the common good and objectivity is the cornerstone of academic efforts, you should be able to work with the interest groups. This would tie many pieces together, would give all the benefit of your expertise, and, not coincidently, provide you the support needed to ensure adequate funding of soil and water research and extension.

This is basically my message but I would like to take a few parting shots.

This workshop and the resulting report should be viewed as a declaration of war—the battles are yet to be fought and won.

Identify your audiences and develop appropriate messages for each.

Recognize that you are treading on thin ice. The arena includes people from the Sagebrush Rebellion to Barry Commoner. You've got to be objective and honest.

Don't throw up your hands because, "they ain't logical". Scientific logic has limited use in the political arena in the short run. Over time, you can make a logical story stick but it takes tenacity.

Be wary of simplistic arguments that we aren't getting a fair share. Al Grable has compiled the soil and water research budgets that he interprets as showing inadequate growth and support. This may be true, however, from 1970 to 1978, soil and water research finding increased 83 percent. In real terms that's better than GNP performance or many other agricultural research areas. You need to show that 1970 levels were inadequate because of neglect during our surplus production years. Focus on the new pressures for production and the resulting need for soil and water research.

Finally, get dirty. You can't stay aloof nor can you pass the buck. Further, you've got to involve top administrators as well as scientists.

SPONSORS

SPONSORING SOCIETIES

American Agricultural Economics Association
American Society for Horticultural Science
American Society of Agricultural Engineers
American Society of Agronomy
Crop Science Society of America
National Association of Conservation Districts
Society for Range Management
Society of American Foresters
Soil Conservation Society of America
Soil Science Society of America

SUPPORTING ORGANIZATIONS

Department of Agriculture (USDA)
- Economics and Statistics Service
- Forest Service
- Science and Education Administrati(
- Soil Conservation Service

Department of the Interior (USDI)
- Bureau of Mines
- Office of Surface Mining
- Office of Water Research and Technology
- Water and Power Resources Service

Farm Foundation
Tennessee Valley Authority
Wildlife Management Institute

OTHER AGENCIES ASSISTING

Argonne National Laboratory
Central and Southwest Services
Clemson University
Colorado State University
Cornell University
Council for Environmental Quality
Humboldt State University
Iowa State University
Kansas State University
Michigan State University
National Agricultural Lands Study
National Association of State Universities and Land Grant Colleges
New Mexico State University
North Carolina State University
North Central States Experiment Stations
North Dakota State University
Ohio State University
Oklahoma Land and Cattle Company
Pennsylvania State University
Purdue University
Randall & Blake, Inc.
Resources for the Future
Southeast Wisconsin Regional Planning Committee
State University of New York
Texas A&M University
U.S. Environmental Protection Agency
U.S. Geological Survey
University of California
University of Florida
University of Georgia
University of Idaho
University of Kentucky
University of Maryland
University of Minnesota
University of Missouri
University of Nebraska
University of Washington
University of Wisconsin
University of Wyoming
Utah State University
Virginia Polytechnic Institute and State University
Washington State University
Welder Wildlife Refuge
Wisconsin Department of Natural Resources

This manuscript is submitted for publication with the understanding that the U.S. government is authorized to reproduce and distribute reprints for governmental purposes.

The views and conclusions contained in this document are those of the authors and should not be interpreted as necessarily representing the official policies, either expressed or implied, of the U.S. government. Any opinions, findings, and conclusions or recommendations expressed in this publication are those of the author(s) and do not necessarily reflect the view of the U.S. Department of Agriculture.

Workshop supported by the U.S. Department of the Interior, Water and Power Resources Service, under Service Agreement No. 1-07-81-V0172.

This material is based on work supported by the U.S. Department of Agriculture under Agreement No. 59-32R6-1-9.

This program is supported by assistance from the Tennessee Valley Authority (TVA), a Federal agency. Under Title VI of the Civil Rights Act of 1964, Section 504 of the Rehabilitation Act of 1973, the Age Discrimination Act of 1975, and applicable TVA regulations at 18 C.F.R. pts. 1302, 1307, and 1309, no person shall, on the grounds of race, color, national origin, handicap, or age, be excluded from participation in, be denied the benefits of, or otherwise be subjected to discrimination under this program. In addition, no qualified handicapped person shall, on the basis of handicap, be subjected to discrimination in employment (including hiring) under the program. If you feel you have been subjected to discrimination as described above, you personally or by a representative, have the right to file a written complaint with TVA not later than 90 days from the date of the alleged discrimination. The complaint should be sent to Tennessee Valley Authority, Office of Equal Opportunity Compliance, 400 Commerce Avenue, Knoxville, Tennessee 37902. A copy of the applicable TVA regulations may be obtained on request by writing TVA at the address given above.

WORKSHOP PARTICIPANTS

Earl Aldon
Rocky Mountain Experiment Station
2205 Columbia S.E.
Alburquerque, NM 87106

David Allee
Department of Agricultural Economics
Cornell University
Ithaca, NY 14850

Raymond R. Allmaras
P.O. Box 370
Columbia Plateau Conservation Research Center
Pendleton, OR 97801

W. D. Anderson
USDA, National Research Economic Division
500 GHI Building
12th Street, S.W.
Washington, DC 20250

Richard W. Arnold
Natural Resource Assessments
P.O. Box 2890, USDA, SCS
Washington, DC 20250

Iris Ballew
National Technical Editing Staff
USDA, SEA-AR, 14th Floor
National Agricultural Laboratory
Beltsville, MD 20705

Harold L. Barrows
USDA, SEA-AR
Building 005, BARC-West
Beltsville, MD 20705

Richard Barrows
Department of Agricultural Economics
University of Wisconsin
1450 Linden Drive
Madison, WI 53706

A. C. Basu
USDA, SCS
P.O. Box 2890
Washington, DC 20013

Sandra S. Batie
Department of Agricultural Economics
Virginia Polytechnic Institute and State University
Blacksburg, VA 24061

Kurt W. Bauer
SE Wisconsin Regional Planning Commission
P.O. Box 769
Waukesha, WI 53187

Marvin T. Beatty
University of Wisconsin
432 North Lake Street
Madison, WI 53706

Robert L. Beck
Argonne National Laboratory
Energy and Environmental Division, Building 8
Argonne, IL 60439

Charles Benbrook
Council for Environmental Quality
722 Jackson Place, N.W.
Washington, DC 20006

William A. Berg
Southern Plains Range Experiment Station
2000—18th Street
Woodward, OK 73801

Anson R. Bertrand
Director, Science and Education Administration
USDA, Room 302-A, Administration Building
Washington, DC 20250

Orville W. Bidwell
Department of Agronomy
Kansas State University
Manhattan, KS 66506

Don H. Boelter
Pacific NW Forest Experiment Station
3200 Jefferson Way
Corvallis, OR 97331

Jack J. Bond
USDA, SEA-PPS
Room 230, Building 005, BARC-West
Beltsville, MD 20705

Nicolaas Bouwes, Sr.
Department of Agricultural Economics
University of Wisconsin
340 Agriculture Hall
Madison, WI 53706

Richard Brammer
Randall and Blake, Inc.
4901 South Windermere
Littleton, CO 80120

Michael F. Brewer
New Executive Office Building
National Agricultural Land Study
722 Jackson Place, NW
Washington, DC 20006

Robert H. Brown
Division of Agricultural Engineering
University of Georgia
Athens, GA 30602

Gary D. Bubenzer
Department of Agricultural Engineering
University of Wisconsin
460 Henry Mall
Madison, WI 53706

Jerry Bucheim
Water Operations and Maintenance
U.S. Department of Interior, Water and
Power Resources Service
P.O. Box 24007
Denver, CO 80225

Earl Burnett
USDA, SEA-AR
P.O. Box 748
Temple, TX 76501

Fee E. Busby, Jr.
Division of Range Science
University of Wyoming
Box 3354, University Station
Laramie, WY 82071

John Carlson
Department of Agricultural Economics
University of Idaho
Moscow, ID 83843

Keith D. Cassel
Department of Soil Science
North Carolina State University
Raleigh, NC 27650

Emery N. Castle
Resources for the Future
1755 Massachusetts Avenue
Washington, DC 20036

Richard B. Corey
Department of Soil Science
University of Wisconsin
Madison, WI 53706

Willie R. Curtis
USDA Forest Service
NE Forest Experiment Station
Berea, KY 40403

James M. Davidson
University of Florida
Gainesville, FL 32611

Clive W. Donoho, Jr.
Director of Research
Ohio Agricultural Research and
Development Center
Ohio State University
Wooster, OH 44691

Orvis P. Engelstad
Division of Agricultural Development
Tennessee Valley Authority
Muscle Shoals, AL 35660

Klaus W. Flach
Soil Conservation Service
USDA, South Building
Washington, DC 20250

Homer Folks
Department of Agronomy
University of Missouri
Columbia, MO 65211

G. R. Foster
Department of Agricultural Engineering
Purdue University
West Lafayette, IN 47907

R. G. Gast
Department of Agronomy
University of Nebraska
Lincoln, NE 68583

Stanley P. Gessel
College of Forest Resources
University of Washington
Seattle, WA 98195

Albert R. Grable
USDA
Building 005, BARC-West
Beltsville, MD 20705

Roy M. Gray
USDA, SCS
P.O. Box 2890
Washington, DC 20013

R. John Hanks
Department of Soils and Biometeorology
Utah State University
Logan, UT 84322

Donald G. Hanway
Department of Agronomy
University of Nebraska
Lincoln, NE 68583

H. P. "Trey" Harbert, III
Central and Southwest Services
2700 One Main Place
Dallas, TX 75250

Henry Hart
Department of Political Science
University of Wisconsin
Madison, WI 53706

Earl Heady
Department of Agricultural Economics
Iowa State University
Ames, IA 50011

Donald W. Hedrick
School of Natural Resources
Humboldt State University
Arcata, CA 95521

Carlton H. Herbel
P.O. Box 698
Las Cruces, NM 88001

R. E. Highfill
Engineering, Technical Development
and Application
USDA, SCS, P.O. Box 2890
Washington, DC 20013

Edward A. Hiler
Department of Agricultural Engineering
Texas A&M University
College Station, TX 77843

James Hite
Department of Agricultural Economics
Clemson University
Clemson, SC 29631

James Hornbeck
Forestry Sciences Laboratory
Concord—Mast Road, Box 640
Durham, NH 03824

Arthur G. Hornsby
Department of Soil Science
University of Florida
Gainesville, FL 32601

Keith A. Huston
Director-at-Large, Ohio Agricultural
Research and Development Center
NC States Experiment Station
Wooster, OH 44691

Laurence R. Jahn
Wildlife Management Institute,
709 Wire Building
1000 Vermont Avenue, NW
Washington, DC 20005

Marvin E. Jensen
USDA, SEA-AR, Room 237
Building 005, BARC-West
Beltsville, MD 20705

Johannsen, Chris
Department of Land, Air, and Water
Res.
University of California
Davis, CA 95616

Howard P. Johnson
Department of Agricultural Engineering
Iowa State University
Ames, IA 50011

Edward T. Kanemasu
765 College Heights Circle
Manhattan, KS 66502

Gordon Kearl
Division of Agricultural Economics
University of Wyoming
Box 3354, University Station
Laramie, WY 82071

Dennis R. Keeney
Department of Soil Science
University of Wisconsin
Madison, WI 53706

Keith A. Kelling
Department of Soil Science
University of Wisconsin
Madison, WI 53706

John F. Kelly
Department of Horticulture
Michigan State University
East Lansing, MI 48824

L. G. King
Department of Agricultural Engineering
Washington State University
Pullman, WA 99164

John Konrad
Chief, Special Studies
Wisconsin Department of Natural Resources
101 South Webster Street
Madison, WI 53702

Ervin J. J. Koos
Route 2, Box 67
Shelby, IA 51570

George J. Kriz
Agricultural Research Service, P.O. Box 5847
North Carolina State University
Raleigh, NC 27650

George W. Langdale
USDA, SEA-AR, P.O. Box 555
Southern Piedmont Research Center
Watkinsville, GA 30677

William H. Langer
U.S. Geological Survey
Mail Stop 720
Reston, VA 22092

William E. Larson
University of Minnesota
201 Soil Science Building
St. Paul, MN 55108

Donald Last
University of Wisconsin
Stevens Point, WI 54481

John Letey
Department of Soil and Environmental Science
University of California
Riverside, CA 92521

Lawrence W. Libby
Department of Agricultural Economics
Michigan State University
East Lansing, MI 48823

Birl Lowery
Department of Soil Science
University of Wisconsin
Madison, WI 53706

Robert F. Lucey
Department of Agronomy
Cornell University
Ithaca, NY 14853

Jerry V. Mannering
Department of Agronomy
Purdue University
West Lafayette, IN 47907

William P. Martin
Department of Soil Science
University of Minnesota
St. Paul, MN 55108

Henry F. Mayland
USDA, SEA-AR, P.O. Box 1109
Northern Plains Research Center
Sidney, MT 59270

Z. B. Mayo
Department of Entomology
University of Nebraska
Lincoln, NE 68583

Ralph J. McCracken
Deputy Chief of Natural Resource Assessment
USDA, SCS, P.O. Box 2890
Washington, DC 20013

Walter F. Megahan
U.S. Forest Service
316 East Myrtle Street
Boise, ID 83702

James R. Meiman
Dean, Graduate School
Colorado State University
Fort Collins, CO 80523

Sigurd W. Melsted
Northern Great Plains Research Station
Box 459 Soils
Mandan, ND 58554

Larry C. Merriam
College of Forestry
University of Minnesota
St. Paul, MN 55108

Arnold Miller
USDA, SEA
Administration Building
Washington, DC 20250

Frederick P. Miller
Department of Agronomy
University of Maryland
College Park, MD 20742

Robert H. Miller
Department of Agronomy
Ohio State University
Columbus, OH 43210

William C. Moldenhauer
Department of Agronomy
Purdue University
West Lafayette, IN 47907

D. David Moyer
Department of Agricultural Economics
University of Wisconsin
Madison, WI 53706

Donald R. Nielsen
Department of Agronomy
New Mexico State University
Las Cruces, NM 88003

Gary R. Nordstrom
Ecological Science, Technical
Development and Application
USDA, SCS, P.O. Box 2890
Washington, DC 20013

Harry E. Olson, Jr.
Executive Program Center
4555 Merrywood Lane
Excelsior, MN 55331

Martin D. Openshaw
American Society of Agronomy
677 South Segoe Road
Madison, WI 53711

Paul G. Orth
18905 SW280 Street
Homestead, FL 33031

Gary Osborne
Department of Soil Science
201 Soil Science Building
St. Paul, MN 55108

James C. Patterson
12821 Baker Drive
Silver Spring, MD 20904

Gary W. Petersen
Pennsylvania State University
University Park, PA 16802

Arthur E. Peterson
Department of Soil Science
University of Wisconsin
Madison, WI 53706

Harry B. Pionke
NE Watershed Research Center
University Park, PA 16802

Dick Piper
Head, Land Utility Section
Water Power and Resources Service
P.O. Box 25007, Denver Federal Center
Denver, CO 80225

Gordon Rausser
Department of Agricultural Economics
University of California
Berkeley, CA 94720

Ken Renard
SW Rangeland Watershed Research
Center
442 East 7th Street
Tucson, AZ 85705

James A. Ryan
Municipal Environmental Research
Laboratory
U.S. Environmental Protection Agency
Cincinnati, OH 45268

R. Neil Sampson
National Association of Conservation Districts
1025 Vermont Avenue, N.W., Room 730
Washington, DC 20005

Max Schnepf
Soil Conservation Society of America
7515 N.E. Ankeny Road
Ankeny, IA 50021

Joseph L. Schuster
Department of Range Science
Texas A&M
College Station, TX 77843

W. D. Seitz
Department of Agricultural Economics
University of Illinois
305 Mumford Hall
Urbana, IL 61801

Thomas N. Shiflet
USDA, SCS
P.O. Box 2890
Washington, DC 20013

Francis H. Siddoway
716 9th Avenue, S.W.
Sidney, MT 59270

R. W. Skaggs
Department of Biological and Agricultural Engineering
Weaver Laboratories
North Carolina State University
Raleigh, NC 27650

Melvin D. Skold
Department of Agricultural Economics
Colorado State University
Fort Collins, CO 80523

Charles M. Smith
USDA, SEA-CR
South Agriculture Building, Room 6418
Washington, DC 20250

Stephen C. Smith
College of Agriculture and Life Science
University of Wisconsin
Madison, WI 53706

Albert E. Smith, Jr.
Georgia Experiment Station
Experiment, GA 30212

J. Herbert Snyder
Water Resources
University of California
Davis, CA 95616

Dale Stansbury
1 Dupont Circle, N.W.
Washington, DC 20036

Matthias Stelly
American Society of Agronomy
677 South Segoe Road
Madison, WI 53711

Bobby A. Stewart
USDA, SEA-AR, P.O. Drawer 10
Conservation and Production Research Laboratory
Bushland, TX 79012

James G. Teer
Welder Wildlife Refuge
P.O. Drawer 1400
Sinton, TX 78387

Grant W. Thomas
Department of Agronomy
University of Kentucky
Lexington, KY 40546

Peter M. Tidd
USDA-SCS
P.O. Box 2890
Washington, DC 20013

David M. Van Doren, Jr.
Ohio Agricultural Research and Development Center
Wooster, OH 44691

Martinus T. Van Genuchten
U.S. Salinity Laboratory
4500 Glenwood Drive
Riverside, CA 92501

Jan Van Schilfgaarde
U.S. Salinity Laboratory
4500 Glenwood Drive
Riverside, CA 92501

Leo M. Walsh
Dean, College of Agriculture
University of Wisconsin
Madison, WI 53706

Gerald B. Welsh
USDA, SEA, BARC-West
Beltsville, MD 20705

Dick Whetsell
Oklahoma Land and Cattle Company
Drawer H
Bartlesville, OK 74003

Edwin H. White
Professor of Forest Soils
SUNY College of Environmental Science and Forestry
Syracuse, NY 13210

Robert J. Willard
Twin Cities Research Center, U.S. Bureau of Mines
5629 Minnehaha Avenue, South
Minneapolis, MN 55417

Robert E. Williams
1025 Vermont Avenue, N.W., Suite 730
Washington, DC 20005

Richard E. Witmer
U.S. Department of Interior, EGS Mail Stop 710
Geological Survey
Reston, VA 22092

ACKNOWLEDGMENTS

The manuscripts in this Proceedings were typed on word-processing equipment that was provided as a courtesy by Wang Laboratories through the help of Ms. Sheila Hartford of Madison, Wisconsin.

The revised copies were produced under the able supervision of Mrs. Arlene Larson and staff.

We wish also to acknowledge the help of David C. Brown and Bill Kelly of the Madison Academic Computing Center, University of Wisconsin, in telecommunication and computer typesetting.

SWRW COMMITTEE ASSIGNMENTS

Program
William C. Moldenhauer

Finance
Robert E. Williams

Implementation
Clive W. Donoho, Jr.
Keith A. Huston

Registration
Arthur E. Peterson

Visual Aids
Birl Lowery

Local Arrangements
Leo M. Walsh
Martin D. Openshaw

Publications
Domenic Fuccillo
Keith A. Kelling
Birl Lowery

CROSS-REVIEW PANEL CHAIRMEN

Resource Base
Grant W. Thomas

Soil Productivity
Donald R. Nielsen

Water Quality/Quantity
James M. Davidson

Socio-Economic Impacts
Albert R. Grable

Energy Constraints
R. Wayne Skaggs

STEERING COMMITTEE

William E. Larson and Leo M. Walsh, Co-Chairmen;

Don H. Boelter; Robert H. Brown; Emery N. Castle; Clive W. Donoho, Jr.; Keith A. Huston; William C. Moldenhauer; Robert E. Williams; Martin D. Openshaw, Executive Director

PLENARY SPEAKERS

Anson R. Bertrand, Director, Science and Education Administration, USDA, Washington, DC

Emery N. Castle, President, Resources for the Future, Washington, DC

Ralph J. McCracken, Deputy Chief, Natural Resource Assessment, Soil Conservation Service, USDA, Washington, DC

R. Neil Sampson, Executive Vice President, National Association of Conservation Districts, Washington, DC

Dale L. Stansbury, Director for Governmental Relations/Agriculture, National Association of State Universities and Land Grant Colleges, Washington, DC

EDITORIAL COMMITTEE

B. A. Stewart and Don H. Boelter, Co-Chairmen;
Iris Ballew; Orville W. Bidwell; Gary D. Bubenzer; Richard B. Corey; Domenic Fuccillo; Max Schnepf